SpringerBriefs in Geography

For further volumes:
http://www.springer.com/series/10050

Anup Saikia

Over-Exploitation of Forests

A Case Study from North East India

 Springer

Anup Saikia
Department of Geography
Gauhati University
Guwahati
India

ISSN 2211-4165   ISSN 2211-4173   (electronic)
ISBN 978-3-319-01407-4   ISBN 978-3-319-01408-1   (eBook)
DOI 10.1007/978-3-319-01408-1
Springer Cham Heidelberg New York Dordrecht London

Library of Congress Control Number: 2013944759

Printed on acid-free paper

Springer is part of Springer Science+Business Media (www.springer.com)

Preface

Northeast India remains in some ways terra incognita. Although this status is slowly changing, in terms of availability and output of the scientific literature, much of these cobwebs still remain. Various factors have contributed to this state of affairs, not least the lack of accessibility within the tract. Tucked away in remote corners of the land known as the Seven Sisters, wildlife sanctuaries, and national parks are a treasure trove being given shabby treatment as their flora and fauna increasingly come under the pressure of the Anthropocene. The forests resources of northeast India amounting to almost a quarter of India's forest acreage are slowly but surely being decimated. Existing structures of protection seem unable to cope with growing human population and their fuel and other requirements. The level of poaching of hapless one-horned rhinos in the Kaziranga National Park underlines the scenario. This book was written with the intention of providing a quick-look at the declining forest cover and the increasing level of forest fragmentation in this global biodiversity hotspot.

Satellite data, courtesy of the USGS's Landsat program, and landscape metrics provide the two core areas around which the work centers. If it is able to stoke further interest on forest loss in northeast India and some positives emerge, I will consider its purpose more than served.

Contents

Chapter 1
Introduction

Abstract Tropical forests have not escaped being transformed into human dominated landscapes the world over. Even biodiversity hotspots have been threatened as forests shrink in qualitative and quantitative terms. India's north eastern region is one such biodiversity hotspot being a part of the Indo-Myanmar global biodiversity hotspot. It is an area with high species endemism that has suffered heavy habitat loss in the face of rapidly increasing human population. While satellite data based biennial forest cover assessments by the Forest Survey of India exist micro level case studies are important to understand forest cover changes and local level dynamics. This study is focused on drivers of forest loss in north east India. It uses the Rani-Garbhanga Reserved Forests in the fringe of Guwahati, the Hamren subdivision of the hill district of Karbi Anglong and the Namdapha National Park in Arunachal Pradesh as sample case studies with which the drivers behind forest loss in north east India are examined.

Keywords North east India • Biodiversity hotspot • Rani-Garbhanga • Hamren • Namdapha

1.1 North East India

All over the world ecosystems have been rapidly transformed in the post-2000 period by human populations through increasingly permanent uses of land (Ellis et al. 2010). Like other ecosystems, tropical forests too have not escaped being transformed as human dominated landscapes expand and forests shrink in qualitative and quantitative terms. Tropical forests perform a variety of services, ecological, economic, social-cultural and aesthetic. In many ways tropical forests are of intrinsic importance to life on earth (Saikia et al. 2013) yet they continue to be decimated at varying rates across the tropics. Most tropical forests are extraordinarily

A. Saikia, *Over-Exploitation of Forests*, SpringerBriefs in Geography, DOI: 10.1007/978-3-319-01408-1_1, © The Author(s) 2014

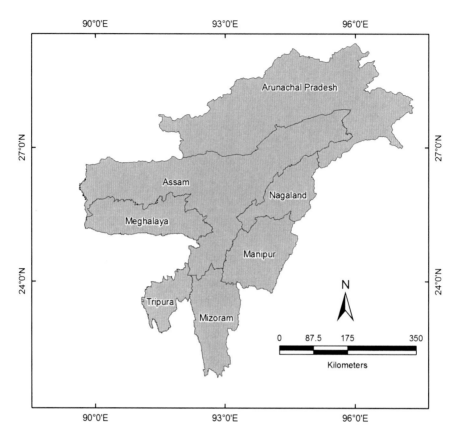

Fig. 1.1 Location map of the NER

rich in species, remarkably complex ecologically and disappearing at truly alarming rates (Laurance 2007). India's north eastern region (NER) (Fig. 1.1) possesses
rich forest tracts that provide various services to some 45 million people of the
region. The NER is one of two of India's mega biodiversity hotspots. Extinction
rates are likely to be higher in biodiversity hotspots which are geographically
restricted areas with high species endemism, heavy habitat loss and rapidly increasing human populations (Laurance 2007). Although the NER is not a geographically restricted area comparable to Madagascar, the Brazilian Atlantic forests or the
Philippines, it is an area with high species endemism that has suffered heavy habitat loss in the face of rapidly increasing human population. Its biodiversity is under
increasing threat in many of its protected areas, reflective of poor enforcement and
protection levels.

The study seeks to analyze land-use/land-cover changes (LUCC) over time in sample areas within the NER and gauge any fragmentation that has occurred to the forest
ecosystem. Finally an assessment of the drivers of forest loss in the NER is made.

1.2 Statement of the Problem

The NER's forest resources are part of the Indo-Myanmar biodiversity hotspot. North east India's forest area accounts for over a quarter of India's forests in terms of areal extent. All this when the NER accounts for a mere 7.7 % of India's total geographical area. However all is not well with the region's forests, which are being degraded and lost at an alarming rate (Sinha 1993, 1996; SFR 1998). The biennial forest cover assessment carried out by the Forest Survey of India (FSI) using satellite images has not been able to present realistic details, especially for north-east India (Roy and Joshi 2002); as such micro level case studies are important to understand forest cover changes.

In many ways, poverty and Hobson's choice are at fault. Poverty, as commonly stated, is the worst polluter and complete lack of choice in seeking alternative avenues of livelihood drives forest loss. There is the need to consider how populations make use of forests, their socio-economic and political organization, the market forces affecting local use patterns and the rules in use related to investment in harvesting of timber and non-timber forest products (Ostrom 1998).

Often the rural economy in many micro-level contexts of the NER represents undiversified economies, with little outside of small scale agriculture. The entire stretch of north east India, barring the Assam valley to a small extent, remains an industrial periphery. The rural non-farm sector as a whole has remained stunted almost ubiquitously across the seven states of north east India. Opportunities to eke out a livelihood are limited and hence there is an over dependence on the forest resources. In hill areas shifting cultivation is naturally closely linked with the forest landscape; and this is important since previous studies have noted the close links between farmers in upland areas and land use/land cover change (Fox 1994).

In considering the drivers of forest loss in the NER, the Rani-Garbhanga Reserved Forests (RGRF) in the fringe of Guwahati, the Hamren sub-division of the hill district of Karbi Anglong and the Namdapha National Park (NNP) in Arunachal Pradesh are sample case studies with which the drivers behind forest loss in north east India are examined.

1.3 Data Used

The study essentially consisted of two segments:

(i) measuring land use/land cover changes in the study sites with particular emphasis to the area under forests
(ii) measuring the degree of fragmentation that has occurred in the study sites using landscape metrics
(iii) assessing the socio-economic and institutional drivers of forest loss.

For the first part of the study, i.e. measuring changes in the forest area, satellite data is to be utilized. Such data has obvious advantages over conventional survey techniques. Multi-date Landsat data was used to assess temporal variations in the areal extent of the forests. Image processing software and geographical information systems were used to process the satellite imageries.

A handheld GPS was used to verify the classifications derived from the satellite imageries. The classifications were analyzed using landscape metrics for the forest areas in particular. Finally personal observation in the field and interactions with stakeholders enabled assessing the drivers of forest loss in the study area.

1.4 Implications

Considering that North East India is a biodiversity hotspot a study relating to forest loss carries implications well beyond the confines of the area itself. Forest losses in biodiversity rich areas are a direct threat to the biodiversity since this translates to habitat fragmentation and raises questions pertaining to the long term survival of endemic species. Satellite data was interpreted digitally using image processing software and the land-use/land cover maps were integrated into a GIS. In addition to land-use/land cover assessments, landscape metrics were examined. Since previous studies (SFR 1998, 2011) have established the issue of forest loss in NER this study looks at the quantum of forest loss in specific sites, their fragmentation and the drivers of forest loss.

1.5 Organization of the Study

The study is organized into eight chapters. This chapter provides an overview including the objectives, datasets used and methods employed and the relevance of the study in the context of north east India, a global biodiversity hotspot. Chapter 2 takes a look at the study area and its physical geography and the status of forests while Chap. 3 conceptualizes the drivers of forest loss in general. Chapter 4 provides details of the data used and the methods followed in the study. Chapter 5 deals with land use/land cover change in the study sites, which are (i) the Hamren sub-division of the hill district of Karbi Anglong in Assam, (ii) the Rani Garbhanga Reserved Forest, also in Assam and (iii) the Namdapha National Park in Arunachal Pradesh. The three study sites present differing institutional frameworks that marks the existing diversity in north east India. The Chap. 6 considers the quantum of fragmentation that has occurred in the three sites detailed in the previous chapter and uses a few landscape metrics to analyze fragmentation of forests in particular. Chapter 7 assesses the drivers of forest loss in relation to the study area while the last chapter provides the conclusions.

References

Ellis EC, Goldewijk KK, Siebert S, Lightman D, Ramankutty N (2010) Anthropogenic transformation of the biomes 1700–2000. Glob Ecol Biogeogr 19:589–606. doi:10.1111/j.1466-8238.2010.00540.x

Fox J (1994) Farmer decision making and spatial variables in Northern Thailand. Environ Manage 18:391–399. doi:10.1007/BF02393868

Laurance WF (2007) Have we overstated the tropical biodiversity crisis. Trends Ecol Evol 22:65–70. doi:10.1016/j.tree.2006.09.014

Ostrom E (1998) The international forestry resources and institutions research program: a methodology for relating human incentives and actions on forest cover and biodiversity. In: Dallmeier F, Comiskey JA (eds) Forest biodiversity in North, Central and South America and the Caribbean: research and monitoring. UNESCO, Paris

Roy PS, Joshi PK (2002) Forest cover assessment in north-east India—the potential of temporal wide swath satellite sensor data (IRS-1C WiFS). Int J Remote Sens 23:4881–4896. doi:10.1080/01431160110114475

Saikia A, Hazarika R, Sahariah D (2013) Land use land cover change and fragmentation in the Nameri Tiger Reserved, India. Geogr Tidsskr Dan J Geogr 113:1. doi:10.1080/00167223.2013.782991

Sinha AC (1993) Beyond the trees, Tigers and Tribes: historical sociology of the Eastern Himalayan forests. Har-Anand, New Delhi

SFR (1996) State of Forest Report 1995. Forest Survey of India, Ministry of Environment & Forests, Government of India, Dehra Dun

SFR (1998) State of Forest Report 1997. Forest Survey of India, Ministry of Environment & Forests, Government of India, Dehra Dun

SFR (2011) State of Forest Report 2011. Forest Survey of India, Ministry of Environment & Forests, Government of India, Dehra Dun

Chapter 2
The Study Area

Abstract The seven states of Arunachal Pradesh, Assam, Manipur, Meghalaya, Mizoram, Nagaland, and Tripura make up north east India. The region spreads over 255,037 km^2 or about 7.7 % of India's total geographical area. The relief is rugged with plateaus, hills and mountains dominating. Level lowlands are at a premium covering only 27 % of the area. Rainfall is torrential with 6,300 mm annually a routine occurrence in the Cherrapunji area. In the rest of the region, average annual precipitation ranges from 1,000 to over 4,000 mm with the bulk of it occurring during the monsoon months of June to October. Stemming from this a luxuriant tropical vegetation ranging from alpine, subtropical pine and montane to evergreen and moist deciduous thrives making the region a global biodiversity hotspot. The focus of development in has centred around its primary resource base, largely around tea, timber and petroleum. What little else exists in the name of development has been concentrated in the urban landscapes. It is not surprising that in the region's rural areas a dependence on natural resources—forests in particular—exists and there is not much option than to exploit the region's forest resources.

Keywords Arunachal Pradesh • Assam • Manipur • Meghalaya • Mizoram • Nagaland • Tripura

2.1 The Land

Spread over an area of 255,037 km^2, the northeastern states of India, commonly referred to as north-eastern region (NER), comprise the seven states of Arunachal Pradesh, Assam, Manipur, Meghalaya, Mizoram, Nagaland, and Tripura. Situated at the north eastern corner of India the area is connected with the rest of the country by a 21 km wide corridor called the "chicken's neck". The NER shares international boundaries with China, Myanmar, Bangladesh and Bhutan.

Generally it has been an area where the level of technology available is on the low side and where humankind has struggled to assert itself; and its peripheral location has been helped. The physiography of the NER, with its imposing elevation, provides the basis upon which human activities are carried out. Given the size of the region and its highly variable terrain there exist not inconsiderable variations in the man-landscape interaction equation. While the landscape has varied from the smooth canvas of the Brahmaputra Valley to the rugged mountainous terrain of the Himalayan Arunachal Pradesh, variations in terms of anthropogenic activities emerged. A brief look at the physiography of the region and the impulses it generated is thus necessary. The NER has been divided into plains, plateaus and hills and mountains (Taher 1986). Low lying plains cover 27 % of the area. These stretches of level land, conducive to settled agriculture with minimal irrigation and other inputs, occur in the Brahmaputra valley which alone accounts for 82 % of the level land in the NER. This stretch of plains covers 56,480 km^2 and runs in a linear manner in a northeast–southwest stretch across the middle of the territory. The Brahmaputra valley plain stretches from Sadiya in the east to Dhrubi in the west and is a near flat stretch of level land built up by the aggradational deposits of the Brahmaputra river, with up to 1,500 m thickness of alluvium. The levelness of the plain is indicated by a fall of about 12 cm/km and the general level ranging from 30 to 130 m (Das et al. 1971) over a length of some 660 km and and an average width of about 70 km (Taher 1986).

Other plains areas include the Barak valley (6,962 km^2), the Manipur Basin (1,843 km^2) and the piedmont plain of Tripura (3,500 km^2) are dwarfed in comparison to the Brahmaputra valley. The Tripura plain, like the Barak is a piedmont plain and represents the northern margin of the Bangladesh plain. Both the Barak and Tripura plains contain erosional as well as depositional features with denuded hillocks and piedmont terraces (Taher 1986). The Manipur plain is the smallest of the plains in the NER and is of lacustrine origin (Taher 1986) probably formed due to the headward erosion of a tributary of the Chindwin river during its incision into the valley, draining out the excess water from the basin, and leaving behind at the deep end, the Loktak lake (Taher 1986). The elevation is less than the 900 m contour that bounds it and this low oval valley has been of seminal importance to the kingdom of Manipur. Like the larger plains, it has a number of local depressions, marshes and lakes, particularly in the south central part (Taher 1986).

Hills and mountains cover an area of some 150,000 km^2 and can be divided into the Arunachal Himalayas in the north; and the eastern hills and mountains consisting of the Dibang-Lohit-Patkai-Naga-Manipur-Mizo axis to the east ranging NE-SW with the Dihang-Dibang gorge being a convenient divisor.

The former is again divided into a Lesser Himalayan zone varying between 300 and 500 m and lying just adjacent to the Brahmaputra valley and to its latitudinal and directional north, the Greater Himalaya, varying in elevation between 5,200 and 7,200 m. The former, a foothills zone or the Sivaliks area, is the more habitable of the two and, given the heavy rainfall it receives, affords a thick, at times impenetrable vegetation. Bands of soft rock, resulting in varying erosion, mean that topographic changes are rapid and the region is generally "a confused labyrinth of hills and ranges intervened by deep gorges" (Taher 1986) with small and marginal valleys scattered in a disorderly manner as if broadcast by some force. It is in such valleys that man is able to eke out a subsistence. Where relatively larger flatlands exist, as in the tablelands of Bomdila and Ziro the traditional shifting agriculture is abandoned for age-old traditional wisdom honed practices of terraced wet rice cultivation. Such enclaves are replaced by the periglacial and inhospitable highlands to the north where, with increase in altitude the vegetation cover peters out, and the snow-capped peaks vie for space with patches of alpine vegetation.

The eastern hills and mountains, aligned NE–SW, are, altitudinally, a poor cousin to the Arunachal Himalayas, ranging from 1,000 to 3,000 m in general and occasionally higher, with the elevation falling off to 150–900 m in Mizoram and Tripura to the south. Northwards, the elevation is the highest at Dapha Bum in Lohit area (4,579 m) gradually decreasing to sub-4,000 m at Saramati (3,826 m) in Nagaland, where the Patkai hills are their best; to Japavo (3,015 m) in Nagaland again in the Barail Range. This decreasing trend continues into Manipur, where the hills enclose the Manipur Central plain in which is embedded the Manipur basin (Taher 1986). Apart from the intermontane basin and a few flat topped valleys like the Khoupam, Manipur is criss-crossed by the Patkai and tributary ranges which trend NE–SW and vary in altitude from 750 m to about 3,000 m, with the highest part of Manipur lying ion the north east at Mount Tenipu (2,994 m), Khayanghung (283 m), Leikot (2,832 m), Siroi (2,568 m) among others.

Plateaus make up the third physiographic division of the NER. Stretching over an area of 32,821 km^2 the Meghalaya-Karbi plateau with an elevation varying between 900 and 1,965 m is a rigid tableland composed of pre-Cambrian Archean gneisses interspersed with Lower Gondwana rocks, Sylhet traps and Cretaceous-Tertiary sedimentaries and the Shillong series (Das et al. 1971).

2.2 Climate and Vegetation

The region's climate is a mixture of cold humid monsoonal climate in hills above 200 m, wet sub-tropical in southern Arunachal, Western Nagaland, Manipur and Mizoram and humid mesothermal monsoonal in the valley and plateau areas (Barthakur 1986). Almost the entire region receives copious rainfall, particularly the Cherrapunji-Mawsynram-Pynursla belt of the southern part of Meghalaya, which borders Bangladesh with a classic scarp face and with its funnel-like topography traps the rain laden winds to receive in excess of some 6,300 mm annually. In the rest of the region, average annual precipitation ranges from 1,000 to over 4,000 mm with about 60 % being concentrated during the months of June to October. This heavy rain, coupled with the hills has clad the region, at least the hills, with a luxuriant vegetation and rich biodiversity. Although the climate has been both a boon to the vegetation which has been much maligned, forests being exploited recklessly, the fact that climate has a profound influence on the life, economy and cultural fabric of the region is undeniable.

Considering the variations in elevation, soil and climate local variations in vegetation are numerous. They can be broadly divided as follows: tropical, deciduous, grasslands, subtropical mixed, temperate and alpine forests (Gopalakrishnan 1991). Of these the tropical forests which include wet evergreen and semi evergreen forests, dry and moist deciduous forests are the single most extensive category covering large stretches of Assam, Meghalaya, Tripura, Mizoram and Manipur. Such forests are dappled with patches of wet bamboo brakes, cane brakes, riparian forests and swamps and pioneer euphorbiaccous serubs. Hollock (Terminalia Myriocarpa), Hollong, Bonsum (Phoebe goalparensis), Nahor (Mesua ferrea), Mekai, Sopa (Magnolia spp.), Tita sopa (Michelia champaca), Simul (Bombax ceiba), Kadam, Makrisal (Schima wallichi), Amari (amoora wallichii) are common trees.

The deciduous forests are spread over parts of Assam, Meghalaya, Tripura and Mizoram. In Meghalaya such forests, in low altitudes of the Khasi and Garo Hills support sal (shorea Robusta) forests (Bhakta 1991), while in the Goalpara, Kamrup, Dhubri, Kokrajhar, Nagaon districts sal species occur, although in limited extent. Other species include Simul (Bombax ceiba), Sidha, Gameri (Gmelina Arborea), Makri-Sal. Evergreen trees jostle for space with the deciduous trees with the former occurring in greater proportion on account of the high rainfall received across most of north east India.

Grassland or savannah vegetation type are commonly found in the lowlands of the Brahmaputra valley which are subject to annual flooding and in area of the Meghalaya Plateau. The original semi-evergreen and deciduous forests have been degraded into grasslands, which represent secondary forests. Vegetation includes grasses, marsh forests and swampy vegetation along with species like Cayera arborea, wrightia tomomtosa, Zizyphus, Randia and 'rata' (Imperata arundinacea) (Das et al. 1971).

Subtropical mixed forests in low elevations (up to 1,500 m) of areas of Arunachal Pradesh and temperate forests in parts of the Meghalaya-Karbi plateau and Naga-Mizo Hills and alpine forests in higher elevations of Purvanchal complete the picture. In the temperate belt species like pine, fir, oak, birth, chestnut, magnolia, maple, cherry, fig, moly and cherry trees occur variously while in the alpine forests restricted to higher elevations in the Arunachal Himalayas between 2,700 and 4,300 m, shrubs, jumpers, pine, silver, fir, dwarf rhododendrous and conifers are found (Prasad 1971) Fig. (2.1).

2.3 Population and Resources

The focus of development in north east India (NEI) has centred around its primary resource base, largely around tea, timber and petroleum. What little else exists in the name of development has been concentrated in the urban

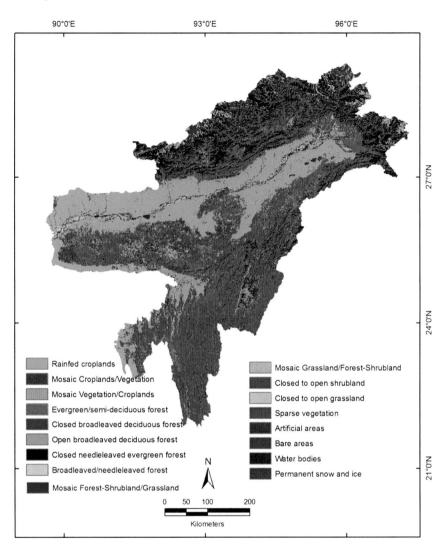

Fig. 2.1 Land use in the NER (© ESA 2010 and UCLouvain)

landscapes. Given such a context it is not surprising that in the region's rural areas a dependence on natural resources (forests in particular) exists. Agriculture and diversification of the rural economy and development of a non-farm sector has developed but this is largely confined to the agriculturally rich lowlands of Assam. Elsewhere there is not much option than to exploiting what remains of the region's forest resources.

Additionally the ethnic multiplicity of the region, frequent clashes between them and militancy have placed environmental priorities low in the list of priorities for the region's state governments. Rapid growth rates of population, the bulk of it attributed to trans-national flows from Bangladesh have not alleviated matters, as an expanding rural population needs a minimum level of fuel wood for subsistence purposes.

To understand land use change in North East India a quick look at its changing population profile seems necessary. Population density is a useful indicator of the form and intensity of human interactions with their ecosystems, as increasing populations have long been considered both a cause and a consequence of ecosystem modification (Ellis and Ramankutty 2008).

Distribution of population in NEI seems closely linked with the physiography. The low lying valleys are densely habited while the hills and plateaus have relatively

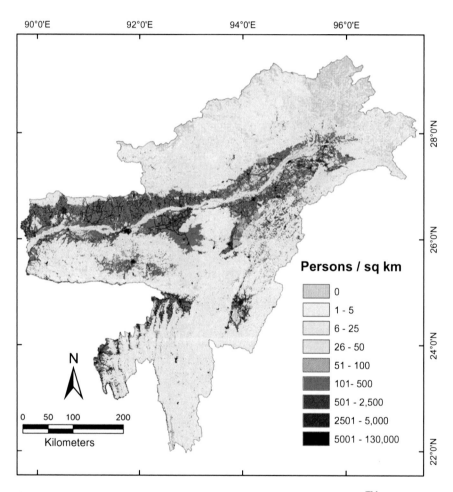

Fig. 2.2 Population density in north east India, 2008 (*Source* Landscan 2008[TM] data provided by www.ornl.gov). The northern-most state, Arunachal Pradesh is the most rugged and mountainous and has the lowest density overall. It possesses snow and ice bound areas in its northern limits as well a few glaciers and densities are in sharp contrast to the Brahmaputra valley that stretches in the middle of the NER ranging NNE-SW

lower densities. With the Brahmaputra river cutting through the territory form north east to south west dominates the landscape of Assam, population densities are dense here at 397 persons/km^2. The other high density patches are in Tripura and in central Manipur, both areas being lowlands drained by smaller rivers. In the rest of the NEI dominated by hills and plateaus densities are far lower. The region's population distribution (Fig. 2.2) reflects these vicissitudes.

Indeed, from a slow growing epidemic-ravaged population prior to 1,900 in a jungle infested backwater, population has climbed consistently in recent decades. In the past century population growth in NEI has been rapid and sustained.

North-east India consists of seven hill states: Arunachal Pradesh, Assam, Manipur, Meghalaya, Mizoram, Nagaland and Tripura. Of these, Assam is the least hilly since much of it lies in the plains of the Brahmaputra river system (Tisdell and Roy 1997). Most of north-east India, however, consists of hills or mountains deeply dissected by rivers and streams owing to uplifting of the land (Tisdell and Roy 1997). Tribal people make up the overwhelming majority of the population of these states, except for Assam, and shifting agriculture and forest resources play a major role in their economic life (Tisdell and Roy 1997).

The rate of growth of all the states of North East India has been—almost without exception—much higher than the all India level (Table 2.1). The decadal rate of increase of population that was markedly above the national aggregate during the 1971–91 period, though it declined in 1991–2001. While the rates of increase for the North East were 35.87 and 27.45 those at the all India level were 24.66 and 23.85 during 1971–81 and 1981–91 respectively.

Without going into the reasons underlying these high growth rates, the effect of a rapid population growth on land-use/land-over change and forest loss/degradation needs to be kept in perspective. In largely rural societies, north east India included, the domestic energy and fuel wood requirement are largely source from free common resources. Thus fuel wood demand and usage escalates as population pressure increases. Second, it has been established by several studies that the fallow period in shifting cultivation declines with population pressure. Additionally there are other effects such as increased grazing cattle pressure and hunting activities (even poaching) as population growth occurs.

Table 2.1   Growth rates of population in North East India

State/region	Decadal growth rate			
	1971–81	1981–91	1991–2001	2001–2011
Arunachal Pradesh	35.15	36.83	26.21	25.92
Assam	36.05	24.24	18.85	16.93
Manipur	32.46	29.29	30.02	18.65
Meghalaya	32.04	32.86	29.94	27.82
Mizoram	48.55	39.70	29.18	22.78
Nagaland	50.05	56.08	64.41	−0.47
Tripura	31.92	34.30	15.74	14.75
India	24.66	23.85	21.34	17.64

Source Census of India, various years

2.4 The Extent of Deforestation

Compared to the rest of the Indian mainland the NER was brought under British rule relatively late in 1826 and transport, commerce and trade were less developed in this region than elsewhere. Although depletion of forests occurred more slowly in the face of limited accessibility, it also meant that government control of the twin scourges of expanding tea plantations established on cleared forest lands and the encroachment of immigrants peasants from present day Bangladesh on forest lands, were much more ineffective (Tucker 1988). Additionally the opening of timber mills during the 1920s along with the grant of long term leases on 'exceptionally favourable terms' to Marwari and Bengali contractors resulted in the commercialisation of the upper Assam forests (Tucker 1988a). The magnitude of forest loss between 1826 and 1950 is difficult to gauge. Thereafter, however, estimates on the extent of forest are available. Estimates by the Central Forestry Commission reveal a loss of 1,763 km^2 of forest area in the NER between 1951 and 1983 (Kashyap 1990).

Between 1989 and 1993 the NER lost 1,418 km^2 of forest while the rest of India recorded an increase of 1,836 km^2 (SFR 1995). With an additional loss of 316 km^2 recorded during 1993–95 (SFR 1997), the magnitude of forest loss over the 6 year period between 1989 and 95 amounted to 1,734 km^2. This closely approximates the loss which occurred over the earlier three decade period pointing to a clear acceleration in the rate of forest depletion. In reality the extent of forest loss was much higher at 6,066 km^2 but was compensated for by forest regeneration in the abandoned shifting cultivation areas.

Since 1989 the State of Forest Reports (SFR) started being prepared biennially by the Forest Survey of India (FSI) under the Ministry of Environment and Forests, Government of India (http://www.fsi.org.in/). A few states of north east India have registered improvements to their area under forest during the period 1989–2009, barring Arunachal Pradesh, Manipur and Nagaland (Table 2.2). The area under forest climbed from 169,379 km^2 in 1989 to 173,780 km^2 in 2009, a

Table 2.2 Changes in forest cover in NER (area in km^2)

	1989	1999	2009	Variation'89 – '09
Arunachal Pradesh	68,763	68,847	67,353	−1,410
Assam	26,058	23,688	27,692	1,634
Manipur	17,885	17,384	17,280	−605
Meghalaya	15,690	15,633	17,321	1,631
Mizoram	18,178	18,338	19,240	1,062
Nagaland	14,356	14,164	13,464	-892
Sikkim	3,124	3,118	3,357	233
Tripura	5,325	5,745	8,073	2,748
NER	169,379	166,917	173,780	4,401
All India	640,134	637,293	690,899	50,765

Source SFR various years

gain of some 4,401 km^2. This meant the NER registered an increase of 2.59 %; over the same period forest cover increased at a substantially higher rate of 7.93 for India as a whole.

The marginal improvement in the region's forest acreage as evidenced from a state level disaggregation of the SFR is one apparent aspect of the story and a more complete understanding is possible when data is further disaggregated.

Before delving further into the matter, a rider needs to be spelt out: that strictly speaking, the SFR assessments of various years are not comparable since different data sets and interpretation methods were followed. Generally coarser data sets were available during the early assessments. Likewise the methodology also differed. Visual interpretation methods employed initially; these were replaced by digital interpretation in the post-2001 assessments; the latter is more accurate, efficient and reliable particularly when supported by sufficient ground truth verification methods, since there exists, quite often, certain features that can be mis-classified or mis-interpreted with digital interpretation methods as well. The scale of interpretation puts a limitation on the mapping and delineation of a geographical feature: for instance, at 1:250,000 scale, the smallest forest cover that could be delineated was 25 ha while at 1:50,000 scale this limit is reduced to 1 ha (SFR 2001). Thus the post 2001 assessments enabled smaller patches of forest and tree canopies (1 – 25 ha in extent), to be detected and mapped (SFR 2001).

Prior to 2001, the assessments used coarser satellite imageries, varying between the 36.25 m IRS-1B to the 80 m Landsat MSS, those after 2001 used better resolution Indian Remote Sensing (IRS) satellite data, namely IRS 1C/1D imageries with 23.5 m resolution. Resolution in this context refers to the minimum spatial size of an object on a satellite imagery that can be correctly identified.

The SFR points out this categorically in stating that a direct comparison with the previous assessment of 1999 was not possible on account of differences in technique and scale (SFR 2001) and that most of the 38,245 km^2 gains recorded in the 2001 assessment could be attributed to technical factors and that only a net increase of 418 km^2 had in fact occurred (SFR 2001).

Thus at least some portion of the encouraging gains determined using state level data for the NER must be watered down as well. Nonetheless a comparison between the SFR assessments over various years is necessary since they are the best and only reliable data source pertaining to India's forest scenario on a consistent and repetitive basis. Further the SFR assessments are far more scientific and reliable than those made by other government agencies that are not always substantiated by field verification. Usually there exist considerable differences between the recorded area under forests (the latter based on returns by the states and annual administrative reports for forest area) and the SFR data.

While the early SFR assessments categorized forests as dense, open and degraded from the 2003 assessment onwards data is available at a more meaningful level of disaggregation and categorization into very dense, moderately dense and open forests. The states of the NER fared moderately well, largely propped

Table 2.3 Forest cover by density categories 2003–09

	2003			2009		
State	VD*	MD	OF	VD	MD	OF
Arunachal Pradesh	13,904	39,604	14,508	20,858	31,556	14,939
Assam	1,684	11,358	14,784	1,461	11,558	14,673
Manipur	720	5,818	10,681	701	5,474	11,105
Meghalaya	168	6,323	10,348	410	9,501	7,410
Mizoram	84	7,404	10,942	134	6,251	12,855
Nagaland	57	5,650	7,902	1,274	4,897	7,293
Tripura	58	4,988	3,047	111	4,770	3,192
Sikkim	458	1,904	900	500	2,161	696
NER	17,133	83,049	73,112	25,449	76,168	72,163

*VD denotes Very Dense, MD Moderately Dense and OF Open Forest
Source SFR (2005, 2009)

up by Arunachal Pradesh (Table 2.3). Excluding Arunachal Pradesh a gain of only 1,362 km^2 was registered in the very dense category.

Inspite of the generally benign picture derived from the recent SFR documents, the fact remains that forests in the NER are under pressure (Arunachalam et al. 2004; Lele and Joshi 2009; Hazarika and Saikia 2013; Saikia et al. 2013). Herein lies the need of a study to assess the drivers of forest loss in the NER.

Acknowledgments The population map of North East India was prepared utilizing the LandScan (2008)™ High Resolution global Population Data Set copyrighted by UT-Battelle, LLC, operator of Oak Ridge National Laboratory under Contract No. DE-AC05-00OR22725 with the United States Department of Energy. The United States Government has certain rights in this Data Set. The United States Government has certain rights in this Data Set. Neither UT-BATTELLE, LLC NOR THE UNITED STATES DEPARTMENT OF ENERGY, NOR ANY OF THEIR EMPLOYEES, MAKES ANY WARRANTY, EXPRESS OR IMPLIED, OR ASSUMES ANY LEGAL LIABILITY OR RESPONSIBILITY FOR THE ACCURACY, COMPLETENESS, OR USEFULNESS OF THE DATA SET.

References

Arunachalam A, Sarmah R, Adhikari D, Majumdar M, Khan ML (2004) Anthropogenic threats and biodiversity conservation in Namdapha nature reserve in the Indian eastern himalayas. Curr Sci 87:447–454
Barthakur M (1986) Weather and climate of north east India. North East Geogr 18:20–27
Bhakta GP (1991) Geography of Meghalaya. Akashi Book Depot, Shillong
Brown D, Schreckenberg K (1998) Shifting cultivators as agents of forest degradation: assessing the evidence. Natural Resource Perspectives Number 29. Overseas Development Institute, London
Das HP, Singh DK, Sharma HN (1971) Meghalaya-Mikir Region In: Singh RL (ed) India: a regional geography. National Geographical Society of India, Varanasi
Ellis EC, Ramankutty N (2008) Putting people in the map: anthropogenic biomes of the world. Front Ecol Environ 6:439–447. doi:10.1890/070062

Gopalakrishnan R (1991) The north east India: land, economy and people. Har-Anand, New Delhi

Hazarika R, Saikia A (2013) The pachyderm and the pixel: an assessment of elephant habitat suitability in Sonitpur, India. Int J Remote Sens 34:5317–5330. doi:10.1080/01431161.2013.787503

Kashyap SC (1990) National forest policy. In: Kashyap SC (ed) National policy studies. Tata McGraw Hill, New Delhi

Lele N, Joshi PK (2009) Analyzing deforestation rates, spatial forest cover changes and identifying critical areas of forest cover changes in north-east India during 1972–1999. Environ Monit Assess 156:159–170. doi:10.1007/s10661-008-0472-6

Prasad J (1971) Eastern himalaya. In: Singh RL (ed) India: a regional geography. National Geographical Society of India, Varanasi

Saikia A, Hazarika R, Sahariah D (2013) Land use land cover change and fragmentation in the Nameri Tiger Reserve, India. Geogr Tidsskr-Dan J Geogr 113:1–10. doi:10.1080/00167223.2013.782991

SFR (1996) State Forest Report 1995. FSI, Ministry of Environment & Forests, Government of India, Dehra Dun

SFR (1998) State Forest Report 1997. FSI, Ministry of Environment & Forests, Government of India, Dehra Dun

SFR (2000) State Forest Report 1999. FSI, Ministry of Environment & Forests, Government of India, Dehra Dun

SFR (2008) State Forest Report 2005. FSI, Ministry of Environment & Forests, Government of India, Dehra Dun

SFR (2009) State of Forest Report 2009. FSI, Ministry of Environment & Forests, Government of India, Dehra Dun

Taher M (1986) Physiographic framework of north east India. North East Geogr 18:1–19

Tisdell C, Roy K (1997) Sustainability of land use in north-east India: issues involving economics, the environment and biodiversity. Int J Soc Econ 24:160–177. doi:10.1108/03068299710161188

Tucker RP (1988) The depletion of India's forests under British imperialism: planters, foresters, and peasants in Assam and Kerala. In: Worster D (ed) The ends of the earth : perspectives on modern environmental history. Cambridge University Press, Cambridge

Tucker RP (1988a) The British empire and India's forest resources: the timberlands of Assam and Kumaon 1914-1950. In: Richards JF, Tucker RP (eds) World deforestation in the twentieth century. Duke University Press, Durham and London

Chapter 3
Conceptualizing Drivers of Forest Loss

Abstract Land-use/land-cover change and forest loss and its drivers have received a significant amount of research attention in recent decades. This attention has transcended discipline specific boundaries and various inert-disciplinary studies in remote sensing, geography, economics, anthropology, landscape ecology, forestry and geographical information systems have addressed the issue of deforestation at widely varying scales. Perhaps at no point of time has the issue of deforestation been more important to humankind then as currently, when climate change and global warming are pressing issues. Among the methods used to assess forest loss and land-use change those using remote sensing data have been found to be accurate and cost-effective. Recent studies have often used landscape metrics in conjunction with satellite data to assess landscape fragmentation. This chapter briefly analyses recent trends in land use/land cover studies.

Keywords Land use/land cover change • Forest loss • Remote sensing • Landscape metrics • Landscape fragmentation

3.1 Conceptualizing Deforestation

If the conceptual and methodological development of a discipline can be compared to that of the branches of a tree, then the multi-disciplinary focus on forest decline can considered as complex; constantly branching out and developing new roots over time. Such an analogy seems legitimate when we consider both the disciplinary focus as well as the methodologies used to explore forest loss and its drivers. The various domains—and this list is only illustrative- of geography, economics, anthropology, landscape ecology, forestry, remote sensing and geographical information systems (GIS) have addressed the issue of deforestation at widely varying scales. Perhaps at no point of time has the issue of deforestation been more important to humankind then as currently, when climate change and global warming are pressing issues that no corner of the globe are free from.

Considering that a wide range of disciplines have and continue to deal with forest loss and land-use/land-cover change, it is not surprising that a wide range of methods are used. Among these that of using remote sensing data which has several advantages in offering a speedy, accurate and cost-effective solution to assessing land-use change (or forest cover change) has been widely used in recent years.

Literature pertaining to the causes of forest loss is substantial and extensive. Basically, the earlier studies have been made at three different levels: household, regional and national including inter-country studies (Kaimowitz and Angelsen 1998). National and regional level studies banking upon secondary data have identified a host of factors responsible for forest loss in different parts of the world. Numerous studies suggest that more forests are cleared if they are closer to roads and market centres. Proximity to roads and railways has also been found to increase the rate of deforestation in Cameroon and Zaire (Mamingi et al. 1996). In addition, areas with high quality soils and drier climates are also likely to be cleared (Chomitz and Gray 1996; Mertens and Lambin 1997). Once fragmented, forests have a higher risk of being lost than forests in large compact areas. Expansion of cropped areas and pastures is a major source of deforestation (Kaimowitz and Angelsen 1998). Logging seems to be a direct source of deforestation in some contexts and an indirect source in others (Kaimowitz and Angelsen 1998). Fuelwood collection is a source of some significant deforestation (Kaimowitz and Angelsen 1998; Puyravaud et al. 2010) though opinion as to whether fuelwood harvesting leads to forest loss or degradation remains divided (Webb and Dhakal 2011). In eastern Bangladesh, a drastic reduction of the shifting cultivation cycle due to high population growth is contributing to significant deforestation and land degradation (Rahman et al. 2012).

The effect of shifting cultivation on forest loss is a contested notion. While some studies are clear regarding its adverse effects (World Bank 1990; Rahman et al. 2012) others hold that the effects of shifting cultivation on forest cover have been overstated (Dick 1991; Colfer and Dudley 1993). Sunderlin (1997) provides a rationale as to why opinion relating to the role of shifting cultivation has been so polarized:

> Those who argue that shifting cultivation is essential for the long term conservation and management of remaining forests have tended to restrict their argument to the traditional, long-fallow shifting cultivation … In contrast, those who argue that shifting cultivation is a threat to forests are actually referring to the short-fallow shifting cultivation.

In many ways, poverty and Hobson's choice are also found to be at fault. Poverty, as commonly stated, is the prime cause and complete lack of choice in seeking alternative avenues of livelihood drives forest loss (Ostrom 1998). Studies also report that weak property rights are responsible for forest loss (Anonymous 2003). When both poverty and weak property right co-exist it is hugely problematic for forested landscapes and ecosystems.

3.2 Proximate and Underlying Drivers

Scholars tend to divide the main drivers underlying forest loss and deforestation into two categories: proximate causes and underlying causes.

> Proximate causes are human activities or immediate actions at the local level, such as agricultural expansion, that originate from intended land use and directly impact forest cover. Underlying driving forces are fundamental social processes, such as human population dynamics or agricultural policies, that underpin the proximate causes and either operate at the local level or have an indirect impact from the national or global level (Geist and Lambin 2002).

In a meta analysis based on 152 cases of tropical deforestation, Geist and Lambin (2001, 2002) identify four broad clusters of proximate causes—agricultural expansion, wood extraction, infrastructure extension and other factors—and five underlying causes—demographic, economic, technological, policy and institutional, and cultural factors.

From tropical deforestation literature, Rudel et al. (2000) identify 20 important factors including timber exploitation and logging, agriculture, fuel wood, roads and so on, that tangent on deforestation.

Several studies using meta-analysis (Kaimowitz and Angelsen 1998; Angelsen and Kaimowitz 1999; Rudel et al. 2000; Geist and Lambin 2001, 2002; van Vliet et al. 2012) at the household level (including other levels) are replete with arguments and counter arguments against the factors causing deforestation.

Forest decline cannot be explained based on a simple causal explanation (Dessie and Kinlund 2007) and a multitude of factors are at play often operating at different scales. While in tropical Australia forest clearing for agriculture and grazing land use is the major disturbance process on private land (McAlpine and Eyre 2002) elsewhere smallholder agricultural expansion and similar anthropogenic factors have impacted landscapes causing habitat loss, fragmentation and forest decline are strongly linked (Dessie and Kinlund 2007; Starzomski and Srivastava 2007; Goodwin and Fahrig 2002; Neel et al. 2004).

In addition to quantifying land-use changes, remote sensing introduced vegetation indices such as NDVI among others that became widely adopted (Runnstrom 2000; Huete et al. 1997). Associated with remote sensing and the use of GIS techniques, (which spread as computing powers mounted and costs declined) are inputs from the discipline of landscape ecology that offered new insights into landscape metrics, patch metrics and various fragmentation indices (Moser et al. 2007; O'Neill et al. 1988).

Of late there has been a tendency among certain studies to combine land-use/land-cover change quantification with understanding patch, class and landscape level metrics (Sivrikaya et al. 2007; Cakir et al. 2007; Sinha and Sharma 2006; Keles et al. 2007; Jha et al. 2000). This is because an understanding of the links between LULC and landscape pattern is pivotal to effective land management and environmental sustainability (Sivrikaya et al. 2007), and quantifying landscape pattern using landscape metrics has been a method of assessing land-use/

land-cover change in several studies (Saikia et al. 2013; Cakir et al. 2007; Lele et al. 2008; Li et al. 2001).

3.3 Summing Up

Land-use/land-cover (LULC) change is considered one of the most important variables of global ecological change (Houet et al. 2010; Vitousek 1994). LULC change affects everything from aerosols and biodiversity to the global carbon and hydrologic cycles (Skole et al. 2004). LULC change can be generated by either natural or human-induced factors or by a combination of the two (Coppin et al. 2004; Mannion 2002).

From the foregoing literature review it can be gauged that:

First, the factors causing deforestation in one area/region may not be applicable to another. Second, while purely bio-physical (Pande and Saha 1994; Roy and Joshi 2002; Jha et al. 2000), historical (Tucker 1988), and socio-economic studies (Saikia 1998) exist, some social scientists have sought to integrate the various drivers of deforestation. Third, although there is a substantial body of literature available at national and sub-national levels, relatively less literature is available at the household–level. Fourth, methods that enabled the combination of land-use/land-cover change quantification with understanding patch, class and landscape level metrics are very useful and have been used successfully in severa contexts.

References

Angelsen A, Kaimowitz D (1999) Rethinking the causes of deforestation: Lessons from economic models. The World Bank Res Observer 14:73–98

Anonymous (2003) ADB forest policy. Working Paper June. Asian Development Bank, Manila

Cakir G, Sivrikaya F, Keles S (2007) Forest cover change and fragmentation using landsat data in Maçka state forest enterprise in Turkey. Environ Monit Assess 137:51–66. doi:10.1007/s10661-007-9728-9

Chomitz KM, Gray DA (1996) Roads, land-use, and deforestation: A spatial model applied to Belize. World Bank Econ Rev 10:487–512. doi:10.1093/wber/10.3.487

Colfer CJP and Dudley RG (1993) Shifting cultivators of Indonesia: Marauders or managers of the forest? Rice production and forest use among the Uma' Jalan of East Kalimantan. Community forestry case study series 6, FAO: Rome. www.fao.org/docrep/006/u9030e/u9030e00.htm Accessed 14 April 2013

Coppin P, Jonckheere I, Nackaerts K, Muys B, Lambin E (2004) Digital change detection methods in ecosystem monitoring: A review. Int J Remote Sens 25:1565–1596. doi:10.1080/0143116031000101675

Dessie G, Kinlund P (2007) Khat expansion and forest decline in Wondo Genet, Ethiopia. Geogr Ann B Hum Geogr 90:187–203. doi:10.1111/j.1468-0467.2008.00286.x

Dick J (1991). Forest land use, forest use zonation, and deforestation in Indonesia: A summary and interpretation of existing information. Background paper to UNCED for the state ministry for population and environment (KLH) and the environmental impact management agency (BAPEDAL)

Geist HJ, Lambin EF (2001). What drives tropical deforestation? A meta-analysis of proximate and underlying causes of deforestation based on sub-national case study evidence. LUCC report series No. 4: Louvain-la-Neuve

Geist HJ, Lambin EF (2002) Proximate causes and underlying driving forces of tropical deforestation. Bioscience 52:143–150. doi:10.1641/0006-3568(2002)052[0143:PCAUDF]2.0.CO;2

Goodwin BJ, Fahrig L (2002) How does landscape structure influence landscape connectivity? Oikos 99:552–570. doi:10.1034/j.1600-0706.2002.11824.x

Houet T, Loveland TR, Hubert-Moy L, Gaucherel C, Napton D, Barnes CA, Sayler K (2010) Exploring subtle land use and land cover changes: A framework for future landscape studies. Landscape Ecol 25:249–266. doi:10.1007/s10980-009-9362-8

Huete AR, Liu HQ, Batchily K, van Leeuwen W (1997) A comparison of vegetation indices over a global set of TM images for EOS-MODIS. Remote Sens Environ 59:440–451. doi:10.1016/S0034-4257(96)00112-5

Jha CS, Dutt CBS, Bawa KS (2000) Deforestation and land use changes in Western Ghats, India. Curr Sci 79:231–238

Kaimowitz D, Angelsen A (1998) Economic models of tropical deforestation: A review. Center for International Forestry Research, Bogor

Keles S, Sivrikaya F, Cakir G (2007) Temporal changes in forest landscape patterns in artvin forest planning unit, Turkey. Environ Monit Assess 129:483–490. doi:101007/s10661-006-9380-9

Lele N, Joshi PK, Agrawal SP (2008) Assessing forest fragmentation in northeastern region (NER) of India using landscape matrices. Ecol Ind 8:657–663. doi:10.1016/j.ecolind.2007.10.002

Li X, Lu L, Cheng GD, & Xiao HL (2001). Quantifying landscape structure of the Heihe River Basin, northwest China using FRAGSTATS. J Arid Environ 48: 521–535 doi:10.1006/jare.2000.0715

Mamingi N, Chomitz KM, Gray DA, Lambin EF (1996) Spatial patterns of deforestation in Cameroon and Zaire. Working paper No. 8, research project on social and environmental consequences of growth-oriented policies, Policies research department, World Bank: Washington, DC

Mannion AM (2002) Dynamic world: Land-cover and land-use change. Arnold, London

McAlpine CA, Eyre TJ (2002) Testing landscape metrics as indicators of habitat loss and fragmentation in continuous eucalypt forests (Queensland, Australia). Landscape Ecol 17:711–728. doi:10.1023/A:1022902907827

Mertens B, Lambin EF (1997) Spatial modelling of deforestation in southern Cameroon: Spatial disaggregation of diverse deforestation processes. Applied Geography 17:143–168. doi:10.1016/S0143-6228(97)00032-5

Moser B, Jaeger JAG, Tappeiner U, Tasser E, Eiselt B (2007) Modification of the effective mesh size for measuring landscape fragmentation to solve the boundary problem. Landscape Ecol 22:447–459. doi:10.1007/s10980-006-9023-0

Neel MC, McGarigal K, Cushman SA (2004) Behavior of class-level landscape metrics across gradients of class aggregation and area. Landscape Ecol 19:435–455. doi:10.1023/B:LAND.0000030521.19856.cb

O'Neill RV, Krummel JR, Gardner RH, Sugihara G, Jackson B, DeAngelis DL, Milne BT, Turner MG, Zygmunt B, Christensen SW, Dale VH, Graham RL (1988) Indices of landscape pattern. Landscape Ecol 3:153–162. doi:10.1007/BF00162741

Ostrom E (1998) 'The international forestry resources and institutions research program: A methodology for relating human incentives and actions on forest cover and biodiversity'. In: Dallmeier F, Comiskey JA (eds.) Forest biodiversity in North, Central and South America and the Caribbean: research and monitoring. 1st edn. UNESCO, Paris

Pande LM, Saha SK (1994) Temporal change monitoring and land-use planning using satellite remote sensing. Asian-Pac Remote Sens J 6:19–29

Puyravaud JP, Davidar P, Laurance WF (2010) Cryptic loss of India's native forests. Science 329:32

Rahman SA, Rahman MF, Sunderland T (2012) Causes and consequences of shifting cultivation and its alternative in the hill tracts of eastern Bangladesh. Agrofor Syst 84:141–155. doi:10.1007/s10457-011-9422-3

Roy PS, Joshi PK (2002) Forest cover assessment in North-East India: The potential of temporal wide swath satellite sensor data (IRS-1C WIFS). Int J Remote Sens 23:4881–4896. doi:10.1080/01431160110114475

Rudel TK, Flesher K, Bates D, Baptista S, Holmgren P (2000) Tropical deforestation literature: Geographical and historical patterns. Unasylva 203:11–18

Runnstrom M (2000) Is northern China winning the battle against desertification? Satellite remote sensing as a tool to study biomass trends in the Ordos plateau in semiarid China. Ambio 29:468–476. doi:10.1579/0044-7447-29.8.468

Saikia A (1998) Shifting cultivation, population and sustainability: The changing context of North East India. Development 41:97–100

Saikia A, Hazarika R, Sahariah D (2013) Land use land cover change and fragmentation in the Nameri Tiger Reserve, India. Geogr Tidsskr-Dan J Geogr 113:1–10. doi:10.1080/00167223.2013.782991

Sinha RK, Sharma A (2006) Landscape level disturbance gradient analysis in Daltonganj south forest division. J Indian Soc Remote Sens 34:234–243. doi:10.1007/BF02990652

Sivrikaya FC, Kadiogullari AI, Keles S, Baskent EZ, Terzioglu S (2007) Evaluating land use/land cover changes and fragmentation in the Camili forest planning unit of north eastern Turkey from 1972 to 2005. Land Degrad Develop 18:383–396. doi:10.1002/ldr.782

Skole DL, Cochrane MA, Matricardi EAT, Chomentowski W, Pedlowski M, Kimble D (2004) Pattern to process in the Amazon region: Measuring forest conversion, regeneration and degradation. In: Gutman G, Janetos AC, Justice CO, Moran EF, Mustard JF, Rindfuss RR, Skole D, Turner-II BL, Cochrane MA (eds) Land change science: Observing, monitoring and understanding trajectories of change on the earth's surface, 1st edn. Kluwer Academic Publishers, Dordrecht

Sunderlin WD (1997) Shifting cultivation and deforestation in Indonesia: Steps toward overcoming confusion in the debate. Rural development forestry network paper 21b:1–18

Starzomski BM, Srivastava DS (2007) Landscape geometry determines community response to disturbance. Oikos 116:690–699. doi:10.1111/j.0030-1299.2007.15547.x

Tucker RP (1988) The British empire and India's forest resources: The timberlands of Assam and Kumaon 1914–1950. In: Richards JF, Tucker RP (eds) World deforestation in the twentieth century. Duke University Press, Durham, pp 91–111

van Vliet N, Mertz O, Heinimann A, Langanke T, Pascual U, Schmook B, Adams C, Schmidt-Vogt D, Messerli P, Leisz S, Castella JC, Jørgensen L, Birch-Thomsen T, Hett C, Bruun TB, Ickowitz A, Vu KC, Fox J, Cramb RA, Padoch C, Dressler W, Ziegler A (2012) Trends, drivers and impacts of changes in swidden cultivation in tropical forest-agriculture frontiers: A global assessment. Glob Environ Change 22:418–429. doi:10.1016/j.gloenvcha.2011.10.009

Vitousek PM (1994) Beyond global warming: Ecology and global change. Ecology 75:1861–1876. doi:10.2307/1941591

Webb EL, Dhakal A (2011) Patterns and drivers of fuelwood collection and tree planting in a Middle Hill watershed of Nepal. Biomass Bioenergy 35:121–132. doi:10.1016/j.biombioe.2010.08.023

World Bank (1990) Indonesia: Sustainable development of forests, land, and water. The World Bank, Washington, DC

Chapter 4
Data and Methods

Abstract Landsat data with a resolution of 30 m at two or more points of time was used to assess land use/land cover change. Since supervised classification necessitates creating numerous training sets, data was collected using a handheld Global Positioning System. A combination of road sampling and transect sampling was followed. During road sampling, points were selected as far as possible at a distance of at least 100 m from the road. This was done to reduce the influence of the road and to avoid generating any bias in the sample that would have occurred had the points are collected close to roads. The maximum likelihood classifier (MLC) which is a widely used classification algorithm was used for the classification and the separability of classes was checked prior to running the classification. Standard accuracy assessment using measures such as user accuracy, producer accuracy and kappa accuracy were performed. Along with Landsat data, ASTER GDEMs and gridded population data from the Landscan dataset of the Oak Ridge National Laboratory for 2008 were used. These datasets were analyzed using Erdas Imagine 9.2 and ArcGIS 9.3 and the spatial statistics program Fragstats 4.1 was used to analyze landscape metrics.

Keywords Landsat • Maximum likelihood classifier • Accuracy assessment • Landscan gridded population data • ASTER GDEM • Landscape metrics

4.1 Introduction

Satellite data was used for a speedy and reliable method to determine land-use/land-cover change (LUCC) in the study area. Landsat data with a resolution of 30 m at two or more points of time was used. Since supervised classification necessitates creating numerous training sets, data was collected using a handheld Global Positioning System (GPS). During fieldwork at various locations in the

Hamren sub-division of Karbi Anglong district and in Rani-Garbhanga Reserved Forest were collected. A combination of road sampling and transect sampling was followed. Road sampling is a fast, partially representative, way of collecting field data. A number of roads in the study area are randomly selected (e.g. by allocating a number to all different roads and then selecting a few of them); sampling points along these roads selected, often by driving along them and stopping the vehicle at a certain intervals (both in terms of distance and/or time). At the stops, sampling points were selected as far as possible at a distance of at least 100 m from the road. This was done to reduce the influence of the road and to avoid generating any bias in the sample that would have occurred had the points are collected close to roads.

Transect sampling is a commonly used sampling design that has been used in the present study. It can be used in areas with no roads, where we had to either travel by foot (e.g. dense forest) or by car (e.g. relatively less accessible conditions). By randomly selecting a starting point for the sampling and then selecting a random direction the bias in selection of ground truth points was reduced.

4.2 Classification of Land-Use/Land-Cover

The maximum likelihood classifier (MLC) which is a widely used classification algorithm was used for the classification. MLC calculates for each class the probability each cell belonging to a particular class depending on its attribute values; with the cell being assigned to the class with the highest probability. All satellite imageries were rectified with the help of Survey of India topographical maps (of 1:50,000 scale) and a common coordinate system was used for the study sites.

Care was taken to take ensure well distributed training sets and the intra-class variabilities were taken into account when selecting clusters of a class, to cover for possible differences in soil conditions and training areas were selected taking different representative pixel samples (Saikia et al. 2013). The number of observations per cluster were kept to a minimum of 30 per band (Janssen and Huurneman 2001) and it was ensured that clusters in the data did not overlap. Following a previous study in the region (Saikia et al. 2013), the separability of classes was checked prior to running the classification. Standard accuracy assessment using measures such as user accuracy, producer accuracy and kappa accuracy were performed. The workflow is represented in the form of a flowchart (Fig. 4.1).

The following satellite datasets were used (Tables 4.1, 4.2, and 4.3):

Along with Landsat data, digital elevation model (DEM) data from the ASTER GDEM and gridded population data from the Landscan dataset from the Oak Ridge National Laboratory (www.ornl.gov/landscan/) for 2008 was used.

These datasets were analyzed using Erdas Imagine 9.2 (ERDAS Inc., Atlanta, GA, USA) and ArcGIS 9.3 (ESRI, Redlands, CA, USA) and the spatial statistics program Fragstats 4.1 (www.umass.edu/landeco/research/fragstats/fragstats.html) was used to analyze landscape metrics.

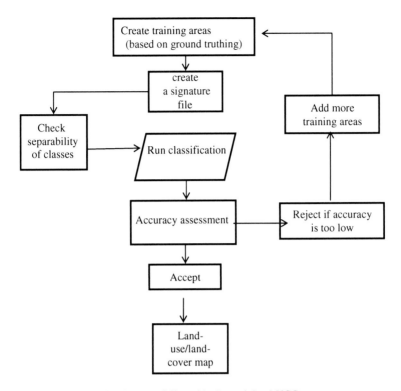

Fig. 4.1 Flow chart showing the steps followed in determining LUCC

Table 4.1 Data used in the LULC classification (Hamren)

Satellite	Number of bands	Resolution (m)	Path/row	Observation date
Landsat 1	4	80	146/42	14 January 1973
Landsat 5	7	30	136/42	26 December 1987
Landsat 7	8	30	136/42	3 December 1999
Landsat 5	7	30	136/42	10 November 2011

Table 4.2 Data used in the LULC classification (Rani-Garbhanga)

Satellite	Number of bands	Resolution (m)	Path/row	Observation date
Landsat 4	7	30	137/42	28 January 1989
Landsat 5	7	30	137/42	11 January 2009

Table 4.3 Data used in the LULC classification (Namdapha National Park)

Satellite	Number of bands	Resolution (m)	Path/row	Observation date
Landsat 5	7	30	134/41	27 October 1988
Landsat 5	7	30	134/41	12 November 2011

References

Janssen LLF, Huurneman GC (2001) Principles of remote sensing. ITC Press, Enschede

Saikia A, Hazarika R, Sahariah D (2013) Land use land cover change and fragmentation in the Nameri Tiger Reserve, India. Geogr Tidsskr-Dan J Geogr 113:1–10. doi:10.1080/00167223.2013.782991

Chapter 5
Land-Use Land-Cover Change

Abstract Land use/land cover changes are occurring at greater pace than ever before in human history. Within the context of land use/land cover change, changes to forest cover are extremely important particularly for rich biodiversity hotspots. This chapter addresses the question: what is the quantum of forest cover loss in the the Hamren sub-division, the Rani-Garbhanga Reserved Forest and the Namdapha National Park. Further forest fragmentation has been assessed using a few indices from landscape metrics.

Keywords Land use/land cover change • Hamren • Rani-Garbhanga • Namdapha National Park • Landscape metrics

5.1 Introduction

The current rates, extents, and intensities of land cover and land use change (LUCC) are far greater than ever before, driving unprecedented changes in ecosystems and environmental processes at local, regional, and global scales (Zhu et al. 2012). Land cover refers to the physical and biological cover over the surface of land, including water, vegetation, bare soil, and/or artificial structures. Land use is characterized by the arrangements, activities, and inputs relating to people in a certain land cover type to produce, change, or maintain it (Zhu et al. 2012).

While land cover is the bio-physical and artificial cover on the earth's surface, including vegetation (natural and planted), as well as man-made constructions, like buildings and roads; land use are the activities to be surveyed, including agricultural operations and operations for nature conservation and environmental management. If land cover consists of 'grassland', then the land use could be 'grazing' or 'recreation'.

Land cover describes the components characterizing the Earth's surface such as vegetation, soils, sediments, water and the built environment; land use reflects the function of land units, notably the human use of land, which often has economic significance (Mannion 2002). Within the context of land use/land cover change, changes to forest cover are extremely important particularly for rich biodiversity hotspots. As such this chapter addresses the question: what is the quantum of forest cover change (loss) in the study area.

The three sites taken up as sample areas are the (i) Hamren sub-division of Karbi Anglong, one of Assam's two hill districts; (ii) the Rani—Garbhanga Reserved Forests (RGRF) on the fringe of the million city of Guwahati; and (iii) the Namdapha National Park (NNP) in the Changlang district of Arunachal Pradesh.

5.2 Land Use/Land Cover Change in Hamren, Karbi Anglong

The Hamren sub-division, like the rest of Karbi Anglong is populated by a predominantly tribal population. Various tribal ethnic groups include the Karbi, Bodo, Kuki, Dimasa, Garo, Tiwa, Hmar and Rengma Naga. Additionally a segment of non-tribal population from neighbouring districts of Assam and other states of North East India and India are settled here. Karbi Anglong with an area of 10,330 Km^2 is the largest district in Assam and easily its richest in terms of flora and fauna. Being a part of the Meghalaya—Karbi plateau the area is rugged and hilly. A population density of 93 persons/km^2 in Karbi Anglong reflects the area's topography; the Hamren density of population is far lower than State of Assam average of 397 person/km^2 (Census 2011). Agriculture is the mainstay of the predominantly rural population. Given the rugged nature of the topography, shifting cultivation (locally known as jhum) rather than settled agriculture is practiced by the Karbi population. In small patches of low-land settled agriculture is carried out however the norm remains jhum cultivation. Agriculture is mostly subsistence oriented reflecting the low level of monetization of the economy. However cultivation of some cash crops such as tea and broom-grass (Thysarolacsa maxzima) is undertaken. With jhum cultivation being the most important agricultural activity, the importance of forests in the Karbi economy is evident. The Karbis harvest the forest for timber, bamboo, cane as well as other minor forest product. Forests and jhum cultivation make up the two pillars of the Hamren economy which is completely bereft of any industrial activity. In the Diphu-Bokajan division to the east of Hamren a pair of cement plants exist.

The Hamren sub-division spreads over an area of approximately 3,105 km^2 (310,587 hectares). In 1973 dense forest accented for 42 % of the Division's area. Along with open forest they accounted for an area of 246,869 hectares or 79 % of the total area of Hamren. During this time grasslands covered only 11 % of the area and non-forest (including settlements and cultivated area) area was nearly 9 % (Fig. 5.1).

The two forest land cover categories (dense forests and open forests) registered a consistent decline from 1973 onwards. By 1987 the combined area under forest had declined to 74 % (230,356 hectares). A sharp decline of these two categories was evident during 1999 as well to 69 % (214,509). A marginal proportional increase was recorded during 1999–2011 to 70.6 % (219,575) (Figs. 5.2, 5.3).

The decline in the two forest land cover categories from 79 % in 1973 to 70.6 % in 2011 by itself is unremarkable. A decline of less than 9 % is hardly significant over a period of 38 years. However, the quantum of loss of forest cover becomes more significant when the changes in the dense forest category are considered (Fig. 5.4).

The area under dense forest declined from 42 % (131,782 ha) in 1973 to 23.46 % (95,188 ha) in 2011 (Tables 5.1, 5.2).

Hamren lost 26,940, 9,653 and 22,285 ha of dense forest between 1973–1987, 1987–1999 and 1999–2011 respectively. A total area of 58878.72 hectares of dense forest was lost during the entire 38 year period.

Dense forest harbours the richest biodiversity and losses of dense forest are virtually irreplaceable as population and human dominated landscapes expand. Unfortunately the loss of dense forest incurred was not merely confined to the community forests but also to the state controlled protected forests in Hamren sub-division. At the eastern corner of the Hamren sub-division lies the Amreng RF. This protected landscape was under the twin forest loss agents of encroachment

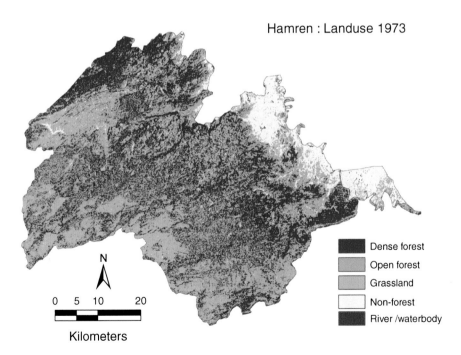

Fig. 5.1   Land use/land cover in the Hamren sub-division of Karbi Anglong during 1973

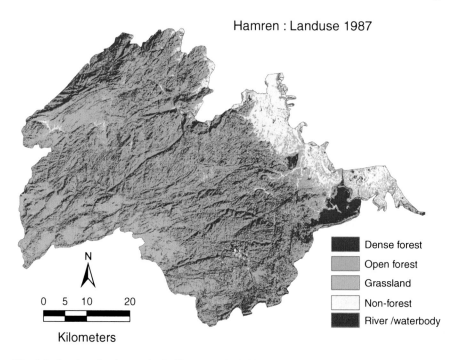

Fig. 5.2   Land use/land cover in the Hamren sub-division of Karbi Anglong during 1987

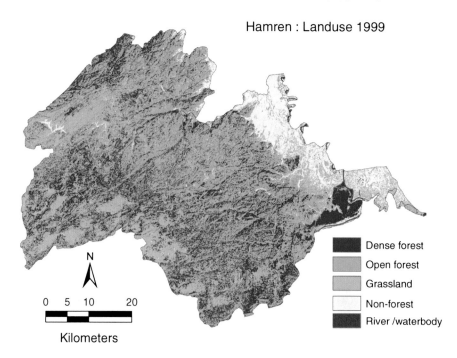

Fig. 5.3   Land use/land cover in the Hamren sub-division of Karbi Anglong during 1999

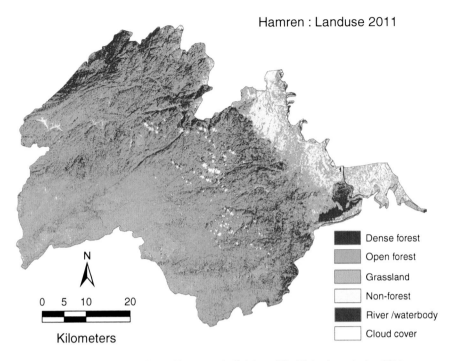

Fig. 5.4 Land use/land cover in the Hamren sub-division of Karbi Anglong during 2011

Table 5.1 Land-use change in the Hamren sub-division (area in hectares)

Land-use	1973	1987	1999	2011	Change (1973–2011)
Dense forest	131782.3	104,842	95188.6	72903.6	−58878.72
Open forest	115086.6	125,511	119320.4	146671	+31584.4
River/waterbody	1796.76	4624.87	1993.5	1776.52	−20.24
Grassland	34977.24	47427.28	65061.2	54346.7	+19369.5
Non forest	27596.52	28182.07	28761.9	27710.5	−113.98
Clouds	–	–	–	7308	–
Total area	311239.4	310587.2	310325.8	310716.6	

Table 5.2 Changes in the dense and open forest categories in Hamren (area in hectares, rounded off)

	1973		1987		1999		2011	
	Area	%	Area	%	Area	%	Area	%
Dense forest	131,782	42.34	104,842	33.76	95,188	30.67	72,903	23.46
Open forest	115,086	36.98	125,511	40.41	119,320	38.45	146,671	47.20
	246,869	79.32	230,353	74.17	214,509	69.12	219,575	70.67

and tree felling (Choudhury 2009) and satellite image data attests to such activity (Figs. 5.5, 5.6 and 5.7).

Annual deforestation rates were calculated using the compound interest rate formula (Puyravaud 2003; Vuohelainen et al. 2012):

$$P = \left[100 \div \left(t^2 - t^1 \right) \right] \ln \left(A^2 / A^1 \right)$$

where P is percentage of forest loss per year, and A1 and A2 are the amount of forest cover at time t^1 and t^2, respectively.

The average rate of deforestation was 1549.44 ha year^{-1} amounting to a rate of loss of 4.75 % year^{-1}. Such a rapid rate of a decline is severely detrimental to the biodiversity of this forest landscape. All the other land-use categories increased with open forests registering the maximum gain of 31584.4 ha in absolute terms. Grasslands increased by 19369.5 ha which was the highest gain in proportionate terms. The gains made by the other categories, grasslands in particular would be largely weaned from dense forests. Likewise degradation of dense forests would add to the increases in open forests.

A previous study centered on the Diphu-Bokajan divisions indicated that losses to the forest cover in Karbi Anglong between 1992 to 2011 were occuring and a concomitant increase in the area under agriculture (Le Moine 2012). This trend

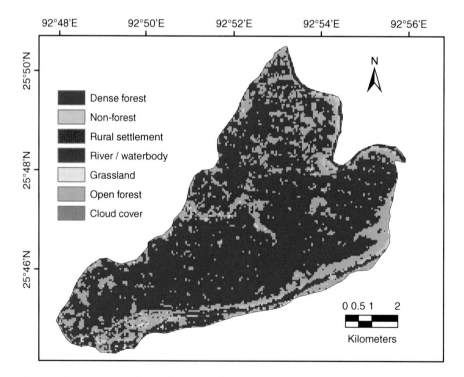

Fig. 5.5 Forest fragmentation in the Amreng RF 1973

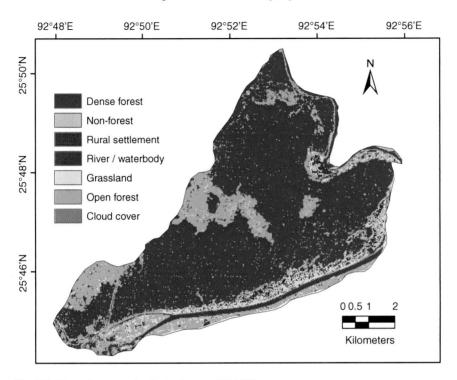

Fig. 5.6 Forest fragmentation in the Amreng RF 1999

seems to apply to the Hamren sub-division as well. Generally large scale jhum cultivation, human settlement (which is designated as encroachment only in RFs) and poaching are problematic in Hamren (Choudhury 2009) along with extraction of firewood (for domestic fuelwood and space heating during winter) and bamboo extraction. The latter is extracted and sold in substantial quantities to the nearby Nagaon Paper Mill at Jagiroad.

Losses to the dense forest land cover category are extremely deleterious to the rich biodiversity that Hamren supports from the hillock gibbon to elephants. Loss of forests are detrimental not only from an reduction in forest acreage but also since fragmentation occurs. The latter adversely affects the mobility of various species within the unconcatenate forest fragments. Fragmentation is evident in the Amreng RF as also in other parts of the Hamren landscape and these will be examined in Chap. 6. A previous study in Karbi Anglong has asserted that conservation measures are crucial and urgently needed (Choudhury 2009) and the present analysis which found a sharp decline in dense forest cover further underlines the need for protection of the existing forest cover (Fig. 5.8).

At the completion of a classification exercise it is necessary to assess the accuracy of the results obtained to allow a degree of confidence to be associated with the results and also to ascertain whether the objectives of the analysis have been achieved (Richards 1993). In fact attribute accuracy is one of

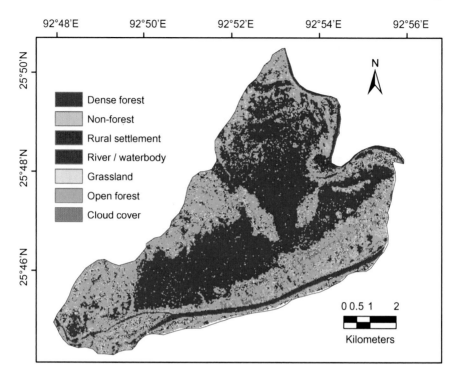

Fig. 5.7   Forest fragmentation in the Amreng RF during 2011

the most critical factors in determining the fitness for use of geographic data and is obtained by comparing values of sample spatial data units with reference data obtained either by field checks or from sources of data with a higher degree of accuracy (Lo and Yeung 2002). In this study field checks was preferred since an accurate map was unavailable. In estimating accuracy sample points were compared with ground truth verification data that was generated using a hand held global positioning system (GPS) device. Sample points were used in view of time, cost and physical accessibility considerations. Accuracy assessment verification using standard measures such as user accuracy, producer accuracy and kappa accuracy were performed to assess the correctness of the classifications.

Accuracy assessments showed encouraging results ranging from an overall accuracy of 88.33 % for 1999 and 90.00 % for 1987 and 2011 (Tables 5.3, 5.4, 5.5); although Kappa accuracy statistics were slightly lower. Since the 1973 imagery was coarser than the remaining imageries and also slightly dated, an accuracy assessment for the 1973 classification was not carried out.

The losses in dense forest in relation to the evaluation as evidenced using a digital evaluation model (DEM) was considered. The Advanced Spaceborne Thermal Emission and Reflection Radiometer (ASTER) Global Digital Elevation Model

Fig. 5.8 The evergreen forests in Hamren, Karbi Anglong. Grazing cattle can be seen at *lower right* corner of the photograph. Paddy cultivation, as distinct from shifting cultivation, is carried out in limited areas, wherever lowlands exist

Table 5.3 Accuracy assessment for the 1987 LULC classification

Class name	Reference totals	Classified totals	Number correct	Producers accuracy	Users accuracy	Kappa accuracy
DF	27	30	27	1	0.9	0.8824
OF	33	30	27	0.8182	0.9	0.8776
Grassland	12	15	12	1	0.8	0.7692
Non forest	12	15	12	1	0.8	0.7692
River/Water body	63	60	54	0.8571	0.9	0.88
Totals	147	150	132			

Overall Classification Accuracy = 90.00 % Overall Kappa Statistics = 0.880

(GDEM) 30 m data was used (Fig. 5.9). The relationship between forest areas and elevation is not clear cut and distinct. The low lying areas to the eastern tip and the north-west have dense forest areas. Conversely the higher elevation area towards the south and south west reveal other land use categories.

Table 5.4 Accuracy assessment for 1999 LULC classification

Class name	Reference totals	Classified totals	Number correct	Producers accuracy	Users accuracy	Kappa accuracy
DF	27	30	27	1	0.9	0.8824
OF	39	30	30	0.7692	1	1
Grassland	14	15	12	0.80	0.8	0.7647
Non forest	13	15	12	0.92	0.9	0.88
River/water body	57	60	51	0.8947	0.85	83.02
Totals	150	150	132			

Overall Classification Accuracy = 88.33 % Overall Kappa Statistics = 0.860

Table 5.5 Accuracy assessment for 2011 LULC classification

Class name	Reference totals	Classified totals	Number correct	Producers accuracy	Users accuracy	Kappa accuracy
DF	33	30	30	0.9091	1	1
OF	27	30	24	0.8889	0.8	0.7647
Grassland	13	15	12	0.8889	0.8	0.7647
Non forest	14	15	11	0.7857	0.7	0.76
River/Water body	57	60	54	0.9474	0.9	0.88
Totals	144	150	132			

Overall Classification Accuracy = 90.00 % Overall Kappa Statistics = 0.80

Thus, there does not seem to exist, as one would have thought, a direct relationship between the higher elevation areas and dense forests. Clearly other factors beyond elevation seem to be at play in determining where dense forests in Hamren remain.

Using a gridded population dataset from the Landscan 2008 data source a visual comparison with the areas of loss of dense forests (Figs. 5.1, 5.2, 5.3 and 5.4) was made. The Landscan gridded population dataset has been used in several studies (Dobson et al. 2000, 2003; Bhaduri et al. 2002, 2007; Rowley et al. 2007).The population count per cell of the Landscan data was converted to persons per km^2 using a geographical information system (GIS) and the population density for Hamren was clipped using the extract by mask function in ArcGIS. The association between high population density areas (Fig. 5.10) and forest loss in the extreme eastern part of Hamren is evident. This area is almost completely devoid of forest cover and is classified as non-forest in all the land use classifications (Figs. 5.1, 5.2, 5.3 and 5.4). The areas with very high density of population (2,501–5,000 and 5,000+ persons per km^2) exist in this part of Hamren. The areas with low density 5–25, 26–50 persons per km^2) occur in the middle part of Hamren and to the west of high density eastern tip of the sub-division. The middle part of Hamren incidentally contains substantial portions of the forest area (dense as well as open forest) of the division. Finally the southern and south-west parts of Hamren indicate moderate to high population density stretches. This area has a

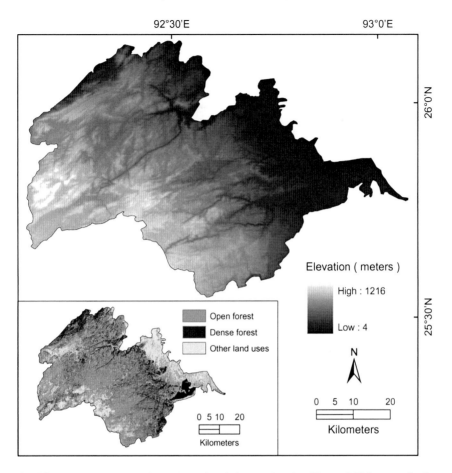

Fig. 5.9 Forest areas (shown in the inset) in relation to elevation (*Source* DEM prepared using ASTER G-DEM, ASTER GDEM is a product of METI and NASA.)

close correspondence between moderate to high population and the lack of forest areas. This area is largely the grassland belt and has little to no forest area. This area has seen the expansion of grasslands since 1973. During 1973 the area had some forest areas interspersed with grasslands. Over the years the share of forest cover began to decline and grasslands began to dominate. It is possible that some open forest conversion to grasslands occurred (Fig. 5.11).

5.2.1 LUCC and NDVI Variability

To explore the effects land-use/land-cover change (LUCC) on vegetation health in the Hamren sub-division, a proxy for the vegetation health namely the Normalized Difference Vegetation Index (NDVI) was used. The NDVI is a widely used

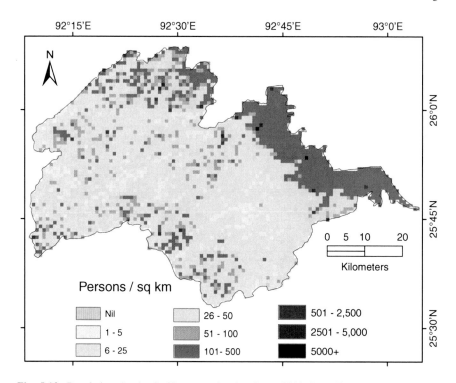

Fig. 5.10 Population density in Hamren using Landscan 2008 data. Figures indicate persons per km[2]

vegetation index and several studies have used it. In this chapter NDVI was derived from the Landsat imageries using the ERDAS Imagine software (ERDAS Inc., Atlanta, GA, USA). The NDVI images obtained for 1973, 1987, 1999 and 2011 indicate the a general decline in vegetation greenness during the period (Figs. 5.12, 5.13, 5.14 and 5.15). In particular the areas in the eastern part of Hamren where settled agriculture and non-forest area dominates the land use, the lowest levels of NDVI exist and this stretch has consistently low greenness values from 1973 through 2011.

In order to derive a more precise estimate of NDVI variability the NDVI images were imported into a geographical information system (GIS); the ArcGIS software (ESRI Inc., Redlands, CA, USA) being used. To derive a spatially aggregate estimate of NDVI (Barbosa et al. 2006) for Hamren 5,000 random points were generated and their NDVI values were extracted using the point locations (Fig. 5.16).

To exclude the non-forest areas to the eastern part of Hamren, the values of 0.40 and below were unselected using the GIS (Fig. 5.17).

A total of 1,316 points were selected with NDVI values above 0.40 during all four years. These represent NDVI values for the dense and open forest categories. 757 pixels with lower NDVI values were representative of grassland and non-forest land use categories (Table 5.6). The remaining random point values of NDVI were not considered as they did not consistently exhibit values above 0.40.

Fig. 5.11   Bamboos being loaded onto a truck for sale at the nearby Nagaon Paper Mill

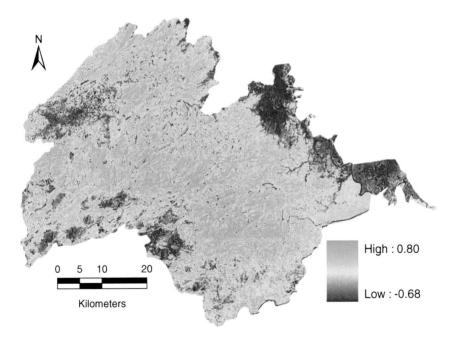

Fig. 5.12   NDVI during 1973

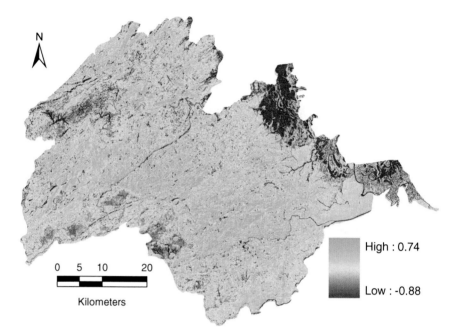

Fig. 5.13   NDVI during 1987

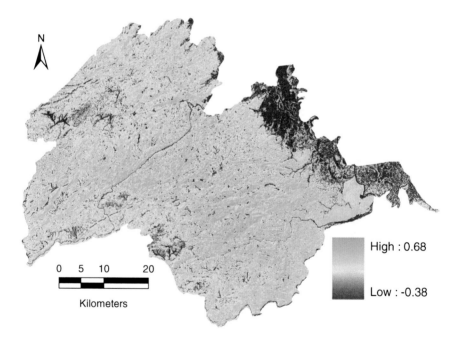

Fig. 5.14   NDVI during 1999

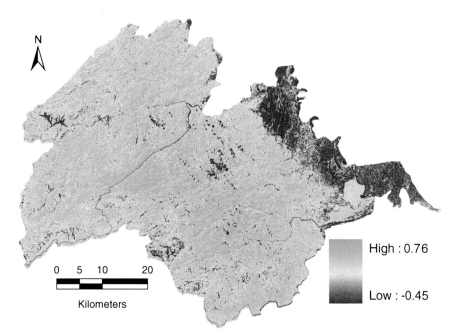

Fig. 5.15   NDVI during 2011

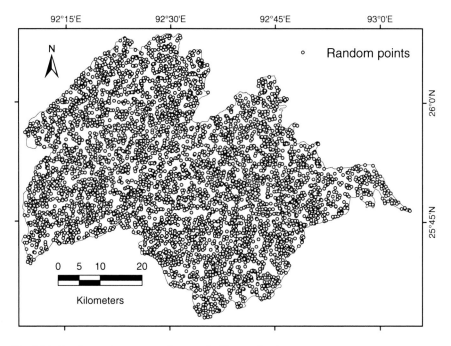

Fig. 5.16   Random points generated across the Hamren landscape

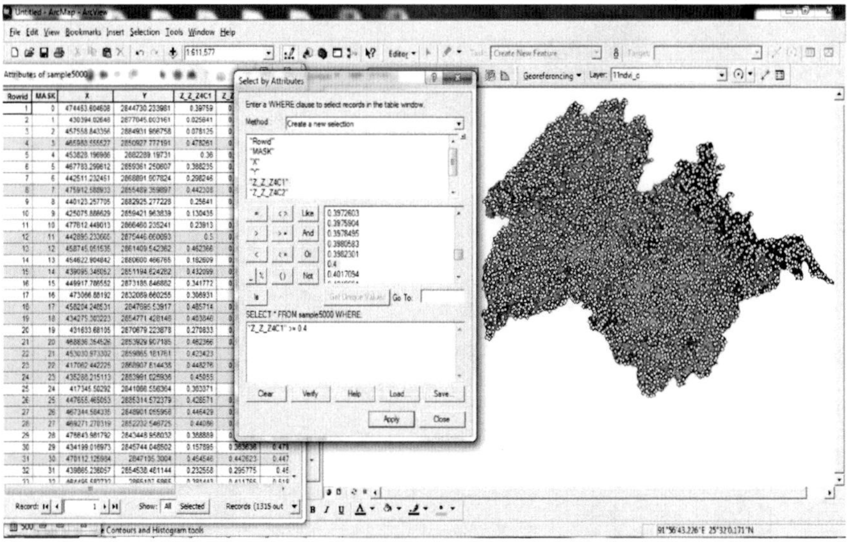

Fig. 5.17 Selecting NDVI by attributes in a GIS

Table 5.6 NDVI changes during 1973–2011

Number of points	1973	1987	1999	2011
757	0.10	0.17	0.23	0.33
1316	0.40	0.41	0.44	0.58

It is evident that NDVI values have increased over the 4 years and it appears that vegetation greenness has not been affected by losses in LUCC; although longer term temporal datasets need to be analyzed to substantiate this. Such a trend is consistent with gains in NDVI made elsewhere, based on time series datasets.

5.3   Land Use/Land Cover Change in the Rani-Garbhanga Reserved Forest

Located on the southern fringe of the million city of Guwahati the Rani Garbhanga Reserved Forest (RGRF) cover an area of 227 km^2. The RGRF consists of two forest ranges, the Garbhanga range (18,361 ha) and the Rani Range (4,369 ha). Sal (Shorea robusta), bamboo (Dendrocalamus hamiltonii), teak (Tectona grandis) plantations, Dipterocarpus macrocarpus, Sehima wallichii, are some of the major species in the RGRF (Barua et al. 2010). The RGRF is prone to human disturbance and activities such as settlements, earth cutting, logging, shifting cultivation at certain places, grazing and road construction (Barua et al. 2010). Among the prominent wildlife species the Asian elephant is found in the RGRF. Some 146 adult

and 189 immature elephants were noted in the RGRF and the nearby Jarasal—
Kawasing forest (Choudhury 1999). The Garbhanga RF supports a rich avifaunal
diversity as well. Some 128 species of birds belonging to 41 families were found
in the RF (Lahkar et al. 2010). To assess changes in the forest cover of the RGRF
imageries of 1989 and 2009 were classified into five broad land use/land cover categories (Table 5.7).

The dense forest land use category (LUC) registered the maximum losses during the 20 year period, losing an area of 1,343 hectares. Moderately dense forest
lost a negligible area of 10.8 hectares. Most of the losses in the dense forest LUC
were converted to non-forest area or open forest/grassland which gained 730 and
658 ha respectively (Fig. 5.18).

Table 5.7 Land-use change in the Rani-Garbhanga RF

Land-use category	1989	2009	Change '89–'09	Gain/loss %
Dense forest	15994.35	14651.01	−1343.34	−8.39
Moderately dense forest	2782.98	2772.18	−10.8	−0.38
Open forest/grass land	1532.25	2190.6	+658.35	+42.96
Non-forest	2321.46	3051.72	+730.26	+31.45
River/waterbody	99.9	65.43	−34.47	−34.50
Total area	22730.94	22730.94		

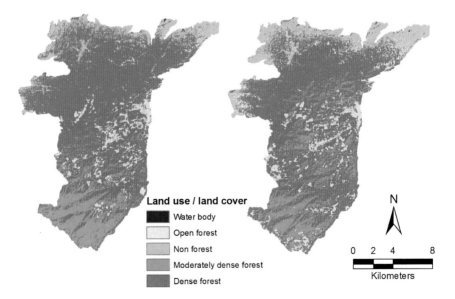

Fig. 5.18 Land use/land cover in the Rani-Garbhanga Reserved Forest during 1989 (*left*) and
2009 (*right*)

Fig. 5.19 NDVI differences
during 1991 and 2005 in
the Rani-Garbhanga RF are
apparent. (*Source* Saikia
2008)

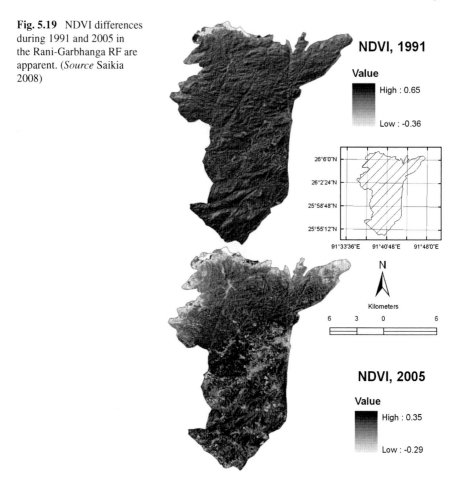

The LUCC losses in the RGRF have also affected the NDVI which has
declined across nearly the entire RGRF landscape and in particular in the northern
fringe, the east-central and southern parts (Fig. 5.19).

Conversion of dense forest into open forest/grassland and non-forest (including cultivated area, homesteads and built-up area) has caused irrevocable damage to this protected landscape's ecosystem. Being located on the southern
fringe of Guwahati city has not helped the cause of the RGRF. The Landscan
2008 population density gridded data (Fig. 5.20) indicates the association
between the higher density pixels in the northern fringes of the RGRF, the
western tip and in the south-central areas with areas of forest loss. While the
northern and western fringe areas have non-forest LUC the south-central parts
possess open forest areas.

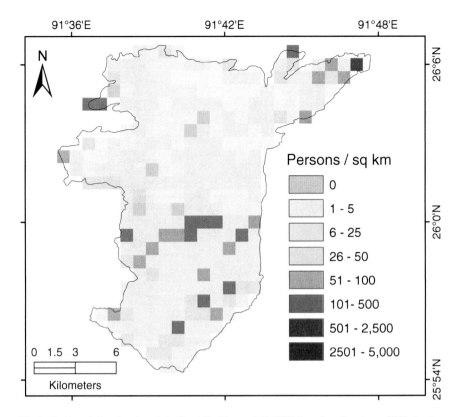

Fig. 5.20   Population density of the Rani Garbhanga RF, 2008 (based on Landscan 2008 data)

5.4 Land Use/Land-Cover Change in the Nandapha National Park

Spread over an area of 1,870 km^2 the Nandapha National Park (NNP) is one of the largest National Park (NP) in India. It is extremely rich in bio-diversity (Datta et al. 2008; Murali Krishna et al. 2012; Arunachalam et al. 2004). However the residents of 25 villages in the buffer areas of the park in the north western and southeastern periphery amounting to a population of some 9,300 persons has adversely affected the virgin forest cover of this park (Arunachalam et al. 2004) (Figs. 5.21, 5.22, 5.23).

Over the 35 years period between 1976–2011 the NNP lost an area of 1,662 hectares of dense forest. Additionally the acreage of open forest/grassland/jhum increased by 1,558 hectares. The gains made by the open forest/grassland areas are evident (Figs. 5.24 and 5.25) in the buffer areas of the park to the east. This is the area where some 25 villages are located and given the nature of the terrain—a low river valley—the increase of open forest at the expense of dense forest is attributable to population pressure (Table 5.8).

Fig. 5.21 A view of the Namdapha National Park. (Photo credit: Ranjan Das)

Fig. 5.22 The Namdapha National Park is one of India's largest parks and possibly the richest in terms of biodiversity. (Photo credit: Ranjan Das)

Fig. 5.23 A woman in one of the 25 villages in Namdapha NP hand-weaves a cloth. The house building materials (largely bamboo, plantain leaves and wooden poles) are all sourced from within the park. (Photo credit: Ranjan Das)

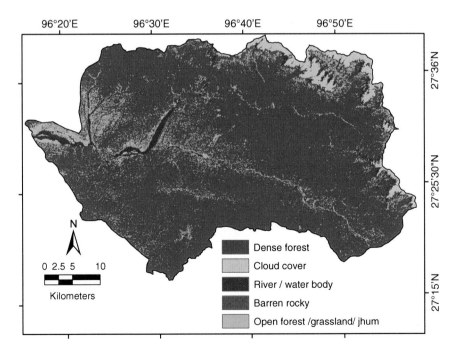

Fig. 5.24 Land use/land cover in the Namdapha National Park during 1976

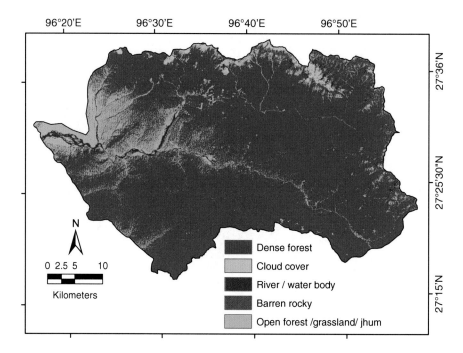

Fig. 5.25 Land use/land cover in the Namdapha National Park during and 2011

Table 5.8 Land-use change in the Namdapha NP

Land-use category	1976	2011	Change 1988–2011	Gain/loss %
Dense forest	157396.50	155734.67	−1661.83	+0.97
Open forest/grassland	17906.77	19464.46	+1557.69	0.31
Barren/rocky area	3633.81	5983.20	+2349.39	64.74
Cloud cover	5006.82	2218.82	−2788.00	61.80
River/water body	3075.12	3617.92	+542.80	9.29
Total area	187019.02	187020.07		

Previous studies have pointed to this trend and how the people of these village extract timber, bamboo, medicinal plants and other non-timber forest products (NTFP) from the NNP (Arunachalam et al. 2004). These effects are clear when the spatial distribution of population in Nandapha is considered (Fig. 5.13) wherein the higher density of 101–500 persons/km^2 is distributed in the eastern river valley segment of the NNP where conversion of dense forests to open forest/grassland/jhum was most distinct (Fig. 5.26).

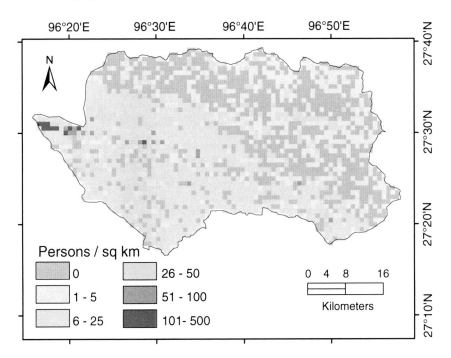

Fig. 5.26 Population distribution in the NNP. Note the higher density in the river valley to the eastern part of the park. This is precisely the area where the conversion of dense forest to open forest/grassland occurred. Extracted using the Landscan 2008 dataset

5.5 Summing Up

A different set of factors has affected the rich forest resources in Hamren, Rani-Garbhanga and Namdapha. The three study sites may be governed by differing institutional frameworks, however the end result has not been radically different: that of forest loss. Hamren in Karbi Anglong, Assam's biodiversity power-house, has been the worst affected as the rate of forest loss has been extremely high amounting to a rate of loss of 4.75 % year-1. Such a rapid rate of a decline is severely detrimental to the biodiversity of this landscape and that the fact that the entire north eastern region of India is part of the Indo-Myanmar biodiversity hot-spot has to be kept in mind in seeking solutions to ameliorate the situation.

Acknowledgements This population density maps for Hamren, Rani-Garbhanga and Namdapha were made utilizing the LandScan (2008)™ High Resolution global Population Data Set copyrighted by UT-Battelle, LLC, operator of Oak Ridge National Laboratory under Contract No. DE-AC05-00OR22725 with the United States Department of Energy. The United States Government has certain rights in this Data Set. Neither UT-BATTELLE, LLC NOR THE UNITED STATES DEPARTMENT OF ENERGY, NOR ANY OF THEIR EMPLOYEES, MAKES ANY WARRANTY, EXPRESS OR IMPLIED, OR ASSUMES ANY LEGAL LIABILITY OR RESPONSIBILITY FOR THE ACCURACY, COMPLETENESS, OR USEFULNESS OF THE DATA SET.

The DEM used in this chapter was prepared using the ASTER GDEM. ASTER GDEM is a product of METI and NASA.

References

Arunachalam A, Sarmah R, Adhikari D, Majumdar M, Khan ML (2004) Anthropogenic threats and biodiversity conservation in Namdapha nature reserve in the Indian Eastern Himalayas. Curr Sci 87:447–454

Barua KK, Slowik J, Bobo KS, Michael Muehlenberg M (2010) Correlations of rainfall and forest type with Papilionid assemblages in Assam in Northeast India. Psyche 560396, 10. doi:10.1155/2010/560396

Barbosa HA, Huete AR, Baethgen WE (2006) A 20-year study of NDVI variability over the Northeast region of Brazil. J Arid Environ 67:288–307. doi:10.1016/j.jaridenv.2006.02.022

Bhaduri B, Bright E, Coleman P, Dobson J. (2002) Landscan: locating people is what matters. Geoinformatics 5:34–37

Bhaduri B, Bright E, Coleman P, Urban ML (2007) LandScan USA: a high-resolution geospatial and temporal modeling approach for population distribution and dynamics. Geo J 69:103–117. doi:10.1007/s10708-007-9105-9

Census of India (2011) Census of India 2011. http://censusindia.gov.in/

Choudhury A (1999) Status and conservation of the Asian Elephant *Elephas Maximus* in North-Eastern India. Mamm Rev 29:141–173. doi:10.1046/j.1365-2907.1999.00045.x

Choudhury A (2009) The distribution, status and conservation of Hoolock Gibbon, Hoolock hoolock, in Karbi Anglong District, Assam, Northeast India. Primate Conserv 24:117–126. doi:10.1896/052.024.0110

Datta A, Anand MO, Naniwadekar R (2008) Empty forests: large carnivore and prey abundance in Namdapha National Park, north-east India. Biol Conserv 141:1429–1435. doi:10.1016/j.biocon.2008.02.022

Dobson J, Bright E, Coleman P, Bhaduri B (2003) LandScan 2000: a new global population geography. In: Mesev V (ed) Remotely-sensed cities. Taylor & Francis, Ltd., London

Dobson J, Bright E, Coleman P, Durfee R, Worley B (2000) LandScan: a global population database for estimating populations at risk. Photogramm Eng Remote Sens 66:849–857

Lahkar D, Lahkar BP, Ahmed F, Talukdar BK, Baruah B (2010) Checklist of the birds of Garbhanga Reserved Forest, Assam, India. Newslett Birdwatchers 50:83–86

Le Moine R (2012) The land use cover changes from 1992 to 2011 in Karbi Anglong, Assam, India. Bachelors dissertation (unpubl.) Linköping University: Linköping

Lo CP, Yeung AKW (2002) Concepts and techniques of geographic information systems. Prentice Hall Inc., New Jersey

Mannion AM (2002) Dynamic world: land-cover and land-use change. Arnold, London

Murali Krishna C, Ray PC, Sarma K, Kumar A (2012) Conservation of White-bellied Heron Ardea insignis (Hume, 1878) habitat in Namdapha National Park, Arunachal Pradesh, India. Curr Sci 102:1092–1093

Puyravaud JP (2003) Standardizing the calculation of the annual rate of deforestation. Forest Ecol Manage, 177:593–596. doi:10.1016/S0378-1127(02)00335-3

Richards JA (1993) Remote sensing digital image processing: an introduction. Springer, Berlin

Rowley RJ, Kostelnick JC, Braaten D, Li X, Meisel J (2007) Risk of rising sea level to population and land area. Eos, Trans Am Geophys Union 88:105–107. doi:10.1029/2007EO090001

Saikia A (2008) Forest fragmentation in North East India. In: Deka S (ed) North East India: geo-environmental issues. Eastern Book House, Guwahati, pp 227–248

Vuohelainen AJ, Coad L, Marthews TR, Malhi Y, Killeen TJ (2012) The effectiveness of contrasting protected areas in preventing deforestation in Madre de Dios, Peru. Environ Manage 50:645–663. doi:10.1007/s00267-012-9901-y

Zhu X, Liang S, Jiang B (2012) Land cover and land use changes. In: Liang S, Li X, Wang J (eds) Advanced remote sensing : terrestrial information extraction and applications. Academic Press, New York

Chapter 6
Landscape Metrics

Abstract Land-use/land cover change also cause habitat loss and fragmentation. Globally habitat loss and fragmentation of landscapes, tropical forests in particular, are the most critical factors contributing to biodiversity losses. From the wide range of landscape metrics that are available to analyze a landscape, three metrics have been used: the number of patches, the patch density and the largest patch index in relation to the Hamren, Rani-Garbhanga and Namdapha landscapes. The analyses indicate that fragmentation is occurring across all three landscapes and that the dense forests in particular have been the most adversely affected. Of the three Hamren has been the most adversely affected while Namdapha has been relatively the least affected.

Keywords Habitat loss • Fragmentation • Number of patches • Patch density • Largest patch index

6.1 Habitat Loss and Fragmentation

The term 'habitat fragmentation' describes the process of transformation from an interconnected into a patchy ecosystem structure (Habel and Zachos 2012). Habitat loss and fragmentation brought on by anthropogenic pressures are the most critical factors contributing to biodiversity losses worldwide (Habel and Zachos 2012; McAlpine and Eyre 2002; Fahrig 1997; Echeverriaa et al. 2006) and the concomitant edge effects that accrue have been a dominant driver of fragment dynamics, strongly affecting forest microclimate, tree mortality, carbon storage, fauna, and other aspects of fragment ecology (Laurance et al. 2011).

Among the numerous negative effects of fragmentation is the reduced capacity of a patch of habitat to sustain a resident population (Iida and Nakashizuka 1995). At forests fragment easier access to remaining forest patches from the edge are provided and gradually detrimental edge effects extend into interior forest areas from these transition zones (Broadbent et al. 2008). A decline in the patch size is additionally problematic since it affects the ability of the patch to support stable or

A. Saikia, *Over-Exploitation of Forests*, SpringerBriefs in Geography,
DOI: 10.1007/978-3-319-01408-1_6, © The Author(s) 2014

productive (source) populations (Nol et al. 2005) of certain species. Species diversity is also adversely impacted by fragmentation (Iida and Nakashizuka 1995). Habitat heterogeneity also declines with fragmentation; and it has been pointed out that habitat aggregation (declining habitat heterogeneity) adversely affects species richness (Steiner and Kohler 2003).

A wide range of landscape metrics are available to analyze a landscape. In this chapter three metrics have been used: the number of patches (NP), the patch density (PD) and the largest patch index (LPI). Landscape metrics can be analyzed at three levels: at patch, class and landscape level. In this analysis class level analysed has been used to identify fragmentation of the dense forest category in particular. The landscape metrics were derived using the programme Fragstats 4.1 (McGarigal et al. 2012). The NP is a basic measure indicative of the degree of fragmentation of a landscape. PD expresses the number of patches on a per unit area basis, i.e. the number of patches per 100 hectares (McGarigal et al. 2012). LPI equals the percentage of the landscape comprised by the largest patch; it approaches 0 when the largest patch of the corresponding patch type is increasingly small and equals 100 when the entire landscape consists of a single patch of the corresponding patch type (McGarigal et al. 2012). At the class level, LPI quantifies the percentage of total landscape area comprised by the largest patch and as such, it is a simple measure of dominance (McGarigal et al. 2012).

6.2 Landscape Metrics in Hamren

In the Hamren sub-division the landscape has fragmented very rapidly during the period 1973–2011 (Table 6.1). The entire landscape across all the land use/land cover categories recorded high levels of fragmentation. More importantly, rapid fragmentation of the dense forest category accrued. At a class level dense forest patches increased exponentially from 4,975 in 1973 to 62,608 by 2011. A similar growth occurred in the grassland category as its NP shot up from 8,229 to 56,625. The rate of increase in the NP was much slower in the case of open forests and non-forest land use categories, with a three and four fold increase respectively.

The increase in the number of patches of the dense forest category is highly correlated (0.975) with time and since 1973 the fragmentation of dense forests has increased in a linear manner ($R^2 = 0.94$) as the NP grew rapidly (Fig. 6.1). Fragmentation of dense forests was evidenced by a rapid rate of increase in the NP. The rate of increase was 532 % during 1973–1987; 632 % during 1973–1999 and 1,158 % during the entire 1973–2011 period. Overall this was an 11-fold increase in the level of fragmentation in the dense forest land use/land cover category; far greater than any other land use/land cover category.

While dense forests increased rapidly, fragmentation of the rest of the landscape did not proceed at a similar pace. In fact the Hamren landscape as a whole registered a 532 % growth in the number of patches or a five-fold increase occurred between 1973 and 2011. If dense forest patches were excluded from the landscape, the degree of fragmentation of the landscape was only four-fold

Table 6.1 Landscape metrics in Hamren

Land use/cover	NP	PD	LPI
1973			
Dense forest	4,975	0.8	17.63
Open forest	13,481	2.17	1.97
Grassland	8,229	1.32	0.88
Non-forest	3,604	0.58	1.47
River/waterbody	1,003	0.16	0.02
1987			
Dense forest	31,448	5.08	1.87
Open forest	34,439	5.56	6.49
Grassland	37,506	6.06	1.36
Non-forest	11,073	1.78	2.55
River/waterbody	15,267	2.46	0.04
1999			
Dense forest	36,456	5.89	1.98
Open forest	63,907	10.33	10.43
Grassland	67,240	10.87	1.17
Non-forest	14,314	2.31	2.61
River/waterbody	5,958	0.96	0.05
2011			
Dense forest	62,608	10.11	1.63
Open forest	39,168	6.32	19.46
Grassland	56,625	9.14	4.48
Non-forest	13,434	2.17	0.98
River/waterbody	2,607	0.42	0.03
Cloud cover	23,633	3.82	0.03

Fig. 6.1 Patch metrics of dense forest

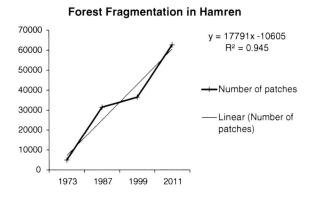

(Table 6.2) in sharp contrast to the 1,158 % increase registered by dense forest land use category. In other words, the rest of the landscape was not as adversely affected by fragmentation as were dense forests.

That dense forests were particularly affected by fragmentation has important connotation for the landscape biodiversity and ecosystem health. Large trees are particularly adversely affected by fragmentation and are unusually vulnerable since they are especially prone to uprooting and breakage near forest edges, where wind turbulence is higher, vines and structural parasites that reduce tree survival

Table 6.2 Patch metrics in Hamren

	Hamren landscape	Dense Forest LUC	Hamren excl. DF
1973	31,292	4,975	26,317
1987	1,29,733	31,448	98,285
1999	1,87,875	36,456	1,51,419
2011	1,98,075	62,608	1,35,467
Rate of growth 1972–2011 (%)	532.99	1158.45	414.75

Fig. 6.2 Patch density and LPI in Hamren

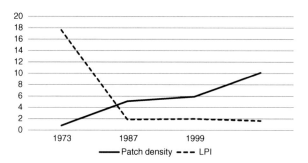

are markedly higher near edges and may be vulnerable to increased desiccation near edges (Laurance et al. 2000). Moreover more trees die near forest edges (Laurance et al. 2000) that are created during fragmentation of forests patches.

Concomitant with the NP of dense forests shooting up, the patch density (PD) increased from 0.8 to 10.11 over the 1973–2011 period, indicative of the dense forest per unit area getting fragmented. PD is inversely related to the largest patch index (LPI) in Hamren. In 1973 high LPI coexisted with a low PD (Fig. 6.2); thereafter as PD began increasing, a consistent decline in LPI began to occur from 1987 through 2011. LPI declined dramatically from 17.63 in 1973 to 1.63 in 2011. The LPI in 2011 was thus only one-ninth of that which existed in 1973. The range of the LPI is from 0 to 100 with index values tending towards 0 indicating very small patches in the landscape and higher values indicating larger patches (McGarigal et al. 2012). In Hamren's dense forests a steep decline in LPI has ecological implications for its biodiversity. Large patches are more conducive to succession and reproduction in certain species and smaller patches can often hinder gene flow among species and migration from one patch to another. For elephants movement between patches can sometimes lead to human elephant conflict, if the inter-patch areas do not have any suitable corridors between them.

6.3 Landscape Metrics in Rani-Garbhanga

For the Rani-Garbhanga Reserved Forest (RGRF) landscape as a whole the NP rose from 7,384 in 1989 to 10,291 in 2009. Likewise the PD increased marginally from 15.53 to 21.65 during the 20 year period. The LPI remained the same at 52.17 (Table 6.3). Within the RGRF landscape however the inter-class changes in

Table 6.3 Landscape metrics in Rani-Garbhanga

Land use/cover	NP	PD	LPI
1989			
Dense forest	1,138	2.39	31.04
Moderately dense forest	2,183	4.59	4.97
Open forest	2,512	5.28	0.42
Non-forest	1,109	2.33	2.35
River/waterbody	442	0.93	0.011
	7,384	15.53	52.17
2009			
Dense forest	1,048	2.2	30.05
Moderately dense forest	3,411	7.17	1.85
Open forest	3,540	7.44	0.3
Non-forest	2,065	4.34	2.71
River/waterbody	227	0.47	0.0098
	10,291	21.65	52.17

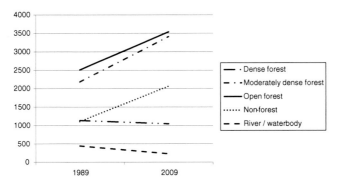

Fig. 6.3 NP by land use category

NP varied considerably (Fig. 6.3). The NP of dense forest declined marginally, as did that for the river/water-body class. Barring these the other classes all registered sharp and substantial increases in NP. The non-forest category showed the maximum increase, as NP nearly doubled in the two decades, reflective of the growing importance of this land use in the RGRF.

The dense forest category showing a marginal decline in the NP does not necessarily mean that forest area improved in any respect since the NP reduced. Instead it resulted because fewer patches of dense forest existed in 2009 than in 1989; when considered in conjunction with a decrease in PD it reflects forest loss in the RGRF.

The other forest categories not only increased in terms of NP but also showed higher PD. Thus the moderately dense and open forest categories had not changed very drastically during 1989–2009 and only nominal changes in these two land-use categories had occurred as seen in Chap. 5. However when the LPI statistic is taken into account clear loss in both moderately dense and open forest categories emerges. LPI declined from 4.97 to 1.85 in moderately dense forest. With LPI already very close to 0 (0.42 in 1989) for the open forest/grassland category, further loss were registered by 2009 taking it to 0.3. The LPI for dense forest showed a marginal decline from 31.04 to 30.05 during 1989–2009. The only category that gained in LPI was the non-forest category which improved its LPI as some consolidation of built-up area seems to have resulted.

6.4 Landscape Metrics in Namdapha

Significant changes in landscape metrics of the Namdapha National Park (NNP) during 1976–2011 occurred. The most important of these pertain to the dense forest and the open forest (including grassland and jhum) categories. Both dense and open forests showing a rapid increase in NP. While dense forests did not decline very significantly, as seen in Chap. 5, a high degree of fragmentation occurred. The NP of dense forests grew from 600 to 3,701 (Table 6.4), an increase of 516 % during 1976–2011; while that of open forest increased from 8,418 to 19,291 by 129 %. The PD of dense forest grew from 0.32 in 1976 to 1.97 by 2011 indicating a greater dispersal of the patches over the NNP landscape.

While the dense forest became far more fragmented, the maximum fragmentation occurred in the western part of the NNP (Figs. 6.4 and 6.5). This area falls in

Table 6.4 Landscape metrics in Namdapha

	1976	2011	1976	2011	1976	2011
Land use/cover	NP		PD		LPI	
Dense forest	600	3,701	0.32	1.98	80.91	82.50
Open forest/grassland/jhum	8,418	19,291	4.50	10.31	0.66	4.18
River	1,234	2,752	0.66	1.47	0.66	1.01
Cloud cover	148	369	2.55	0.20	2.61	0.44
Barren/rocky	4,768	2,749	2.55	1.47	0.53	0.41

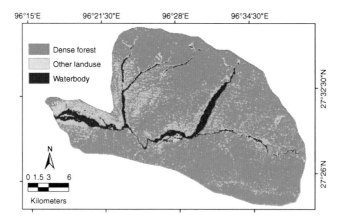

Fig. 6.4 Fragmentation in the western part of Namdapha showing changes during 1976

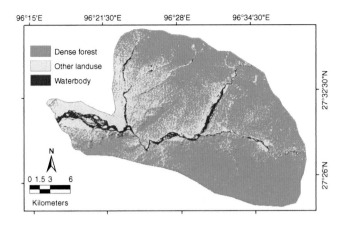

Fig. 6.5 Fragmentation in the western part of Namdapha showing changes during 2011

the buffer area of the NNP where the 25 odd villages are located. Interestingly the LPI made a marginal gain from 80.91 in 1976 to 82.5 in 2011. This would mean that the core area has remained intact and largely unaffected by fragmentation and the LPI could register a slight improvement. That the western part of the NNP has been rapidly fragmented, however, does not augur well for the future.

6.5 Forest Fragmentation

Across all the three landscapes namely Hamren, Rani-Garbhanga and Namdapha forest fragmentation has occurred. This has several deleterious effects on the flora and fauna of these forest ecosystems. When forest patches and grasslands have been lost, degraded and fragmented the ungulate populations tend to decline both in abundance and distribution (Suchitra Devi et al. 2011) and this directly affects higher mammals such as tigers. Namdapha's forest fragmentation is particularly important since is a rich bio-reserve. Habitat fragmentation is one of the main reasons behind the Asian elephant's decline (Sukumar 1989) and Namdapha is an important elephant habitat. The same concern holds valid for the Hamren sub-division and Rani-Garbhanga RF with respect to elephants in particular and the other species in general. That protected areas are losing their rich bio-resources is an alarming issue that must be addressed by the powers that be expeditiously.

References

Broadbent EN, Asner GP, Keller M, Knapp DE, Oliveira PJC, Silva JN (2008) Forest fragmentation and edge effects from deforestation and selective logging in the Brazilian Amazon. Biol Conserv 141:1745–1757. doi:10.1016/j.biocon.2008.04.024

Echeverriaa C, Coomesa D, Salasc J, Rey-Benayasd JM, Larab A, Newtone A (2006) Rapid deforestation and fragmentation of Chilean Temperate Forests. Biol Conserv 130(48):1–494. doi:10.1016/j.biocon.2006.01.017

Fahrig L (1997) Relative affects of habitat loss and fragmentation on population extinction. J Wildl Manag 61:603–610

Habel JC, Zachos FE (2012) Habitat fragmentation versus fragmented habitats. Biodivers Conserv 21:2987–2990. doi:10.1007/s10531-012-0349-4

Iida S, Nakashizuka T (1995) Forest fragmentation and its effect on species diversity in sub-urban coppice forests in Japan. For Ecol Manage 73:197–210. doi:10.1016/0378-1127(94)03484-E

Laurance WF, Delamonica P, Laurance SG, Vasconcelos HL, Lovejoy TE (2000) Conservation: rainforest fragmentation kills big trees. Nature 404:836. doi:10.1038/35009032

Laurance WF, Camargo JLC, Luizão RCC, Laurance SG, Pimm SL, Bruna EM, Stouffer PC, Williamson GB, Benítez-Malvido J, Vasconcelos HL, Van Houtan KS, Zartman CE, Boyle SA, Didham RK, Andrade A, Lovejoy TE (2011) The fate of Amazonian forest fragments: a 32-year investigation. Biol Conserv 144:56–67. doi:10.1016/j.biocon.2010.09.021

McAlpine CA, Eyre TJ (2002) Testing landscape metrics as indicators of habitat loss and fragmentation in continuous eucalypt forests (Queensland, Australia). Landscape Ecol 17:711–728

McGarigal K, Cushman SA, Ene E (2012). FRAGSTATS v4: spatial pattern analysis program for categorical and continuous maps. Amherst: computer software program produced by the authors at the University of Massachusetts. Retrieved from www.umass.edu/landeco/research /fragstats/fragstats.html

Nol E, Francis CM, Burke DM (2005) Using distance from putative source wood-lots to predict occurrence of forest birds in putative sinks. Conserv Biol 19:836–844. doi:10.1111/j.1523-1739.2005.00367.x

Steiner NC, Kohler W (2003) Effects of landscape patterns on species richness – a modelling approach. Agric Ecosyst Environ 98:353–361. doi:10.1016/S0167-8809(03)00095-1

Suchitra Devi H, Hmingthangpuii, Sarma KK (2011) Change in vegetation cover of Dampa Tiger Reserve, Mizoram, North East India: a serious threat to tiger population. J Exp Sci 2:1–6

Sukumar R (1989) The Asian elephant: ecology and management. Cambridge studies in applied ecology and resource management. Cambridge University Press, Cambridge

Chapter 7
Drivers of Forest Loss

Abstract A diverse set of drivers is at play in north east India in affecting forest loss. Growing population, declining forest resources, limited options outside of agriculture and in the hills, an over dependence on forests resources including fuelwood consumption, logging and encroachment. At the same time the onus culpability of deforesting the region should not be conveniently foisted upon shifting cultivators alone. In turn jhum cultivation has been affected by reduced fallow periods as population pressures mount. In toto, the end result has been that this biodiversity hotspot is ending up on the short end of the stick.

Keywords Jhum cultivation • Encroachment • Logging • Population pressure Fuelwood consumption

7.1 Jhum Cultivation

Shifting cultivators are often seen as the primary agents of deforestation in developing countries (Myers 1992) although the role of this traditional method of cultivation is sometimes overemphasised (Angelsen 1995). In the NER as well, shifting cultivation is considered as one of the foremost agents of deforestation (Ranjan and Upadhyay 1999; Sarma 1987; Rajimwale 1991; SFR 1991; Roy and Joshi 2002). In the hill areas of north-east India, shifting cultivation (locally known as jhum) is the main source of livelihood of the people and this is hardly surprising considering the low level of socio-cultural, technology and the negligible proportion of level land available in the state. From the point of view of forests, however, the effects of jhum cultivation are hardly beneficial. In the past when population densities were substantially lower the jhum field/plot was allowed to lie fallow for long years; long enough to allow vegetal cover to regenerate before the cultivators returned to cultivate that plot again and thus complete the jhum cycle. Such long cycles, which were ecologically healthy, soon gave way to reduced cycles once the factor of population and land scarcity came into play. Short cycles are a contrast to the pre-1950s when cycles as long as 20–30 years prevailed and were ecologically sustainable. Current practices are recognised as being neither ecologically nor economically viable (Ramakrishnan 1985). Negative ecological effects result in two ways: first, when shifting cultivators shift their cultivation plots, fresh forest lands need to be cleared and with reduced cultivation cycles that vary from 3–5 years, in the face of increased population pressure on land resources, abandoned plots rarely get enough time to regenerate and second, the secondary forests that grow are much poorer in terms of crown density, species diversity and richness vis-à-vis the primary forests.

While it is easy to label shifting cultivation as a major cause of forest loss, the fact remains that for the shifting cultivator there are many advantages:

> the jhoom fire quickly renders a dense forest fit for growing crops. Fire is thus a great labour saving device and clears the cut jungle in a very short time. The ashes correct the soil activity, and the admixture of ashes with the soil makes the soil more fertile. The soil of hill slopes of high rain zones are generally acidic, which are partly neutralized by the alkali content of the ashes. Further fire clears the areas of extensive preponderance of fungi, insects and pests…the clearing of insects and pests from the surrounding area ….creates a protective belt and keeps the crop safe from the heavy inroads (made by) insects and pests. Burning also retards the growth of tubers and seeds of weeds…. If the fire had not helped him in this manner, he would have laboured more and reaped less. (Barthakur 1977)

While the impact of forest degradation is not all that apparent in a given year, over time as more and more areas are brought under the jhum cultivation practice, the deleterious effects on forest cover, soil loss and soil erosion get compounded. However shifting cultivation is not completely without benefits. The fallow period in shifting cultivation allows soil to stabilize and vegetation to regrow; it also promotes greater carbon sequestration and biodiversity conservation than that in permanent agriculture, as Fox (2000) observes, though it adversely affect species loss through forest fragmentation and also the structure and composition of forests.

7.2 Forest Loss: Beyond Shifting Cultivation

Research suggests that explanations for deforestation must be sought in a variety of factors, many of which should be placed at the door of governments and international capital rather than of shifting cultivators (Brown and Schreckenberg 1998). Although shifting cultivation is identified as the single largest factor behind loss of forest cover in the NER (SFR 1997), factors such as encroachment of forest areas, spread of smallholder agriculturalists, timber consumption within the region and logging in forest areas are also drivers of forest loss.

7.2.1 Encroachment of Forest Areas

Encroachment occurs in forest areas that are perceived as being common resources. Poverty, landlessness and increasing population inflows from neighbouring areas, people displaced by floods or by river bank erosion are all contributing factors.

Table 7.1 Encroached areas in India's National Parks, 2007

State/Country	National park	Encroached area (ha)
Assam	Nameri National Park	2100.00
	Rajiv Gandhi Orang National Park	800.00
	Manas National Park	1700.00
	Dibru-Saikhowa National Park	300.00
	Kaziranga National Park (incl. additional area)	7790.00
Arunachal Pradesh	Namdapha National Park	3005.00
Rest of India		1443.69
India Total		17138.69

Frequently protected areas with low levels of enforcement are prone to encroachment in the NER. Encroachment and poaching by vested interests is occurring in National Parks such as Kaziranga (Das 2006; Levy and Scott-Clark 2007). Encroachment has also been an agent of forest loss and degradation in the Nameri National Park (Saikia et al. 2013). The quantum of protected area under encroachment in the NER is not insubstantial. Out of a total area of 17138.69 ha encroached in some 20 National Parks (NPs) of India the bulk of it occurred in two states of north east India, namely Assam and Arunachal Pradesh (Saikia 2008). The encroached area in Assam amounted to 12,690 ha or 74 % of the nation's total, though its proportionate share of India's geographical area was a mere 2.3 %. This was closely followed by Arunachal Pradesh and these two states left the rest of India, taken together, lagging far behind (Table 7.1).

In 2001 the Ministry of Environment and Forests (MoEF) directed the Assam government to ensure completion of operations against encroachments (Anon 2002); later in 2002 the Supreme Court directed 9 states of India, including Assam, to list steps that had been taken to prevent encroachments on forest land, in particular in hilly terrains, national parks and sanctuaries (Anon 2002). In spite of such efforts encroachments continue. In the Kaziranga National Park a large number of immigrants, are allegedly among the encroachers; though evictions began on a small scale before the drive was 'temporarily' suspended (Anon 2002); a later investigative report stated that these encroachers were from Bangladesh (Levy and Scott-Clark 2007).

7.2.2 Timber Consumption Within the Region

During the pre-1996 period Assam produced 65 % of India's plywood. Four states, Arunachal Pradesh, Assam, Meghalaya and Nagaland all from the NER produced 75 % of India's plywood output (Singh 1990). During about the same period, a substantial timber trade from the NER to the Indian mainland via rivers and railways was going on. During 1993–1994, 71,430 tons or 85 % of the country's trade in teakwood was met from timber originating predominantly in Assam and to a lesser extent from Nagaland and Tripura (Saikia 1998). In the case of 'other timber' (a category for which statistics was collected by the Directorate General of Commercial Intelligence and Statistics) as well, a similar situation existed. Assam was alone despatching two-thirds of the total quantity involved in the national trade in other timber. Along with Nagaland, Tripura and Meghalaya this figure amounted to 73 % of the country's national intra-regional trade in other timber. Much of the teak from the NER found their way to international markets via ports in Gujarat (Saikia 1998). Additionally scores of truckloads would be transported by trucks across the Srirampur border on a daily basis. However the Supreme Court order of December 1996 brought an end to these practices although considerable damage to the region's forest resources had already been made.

The Supreme Court ruling made also put an end to the operation of veneer and saw mills in the NER. While only the state forest departments were permitted to extract timber towards official requirements (Nongbri 2001); illegal trade and covert saw mills, however, continued to operate. Urban markets and those in numerous small towns and rural hamlets operate as makeshift shops to extract valuable timber that is processed and easily sold as furniture. While these activities continue exert an adverse influence on the NER's forest resources, the fact that the levels have declined from the pre-1996 period of timber trade is undeniable.

Timber including sawn timber and logs are the main source of fuel in most of the region's rural landscape. The issue of forest encroachment is closely linked with small agriculturalists who adopt a convenient method to occupy forest land. The encroachers clear the forest over a small area; then set up homestead and an agricultural plot. Once under cultivation it is easily claimed that the homestead and plot has existed for several years, thereby strengthening the illegitimate claim over that land. The practice of encroacher-turned-small-agriculturalist reflects the lack of livelihood alternatives outside of agriculture and poor levels of forest protection and enforcement by institutional authorities. In the fringes of protected areas smallholder agriculture has tended to expand settlements (often illegally) into the forest areas and take a toll on the forest. To supplement their incomes, such encroachers often sell timber in small bundles along the roadside for fuelwood. The result is clear in terms of forest loss, even in protected areas and national parks, as forest extents diminish and fragmentation of forest habitats occur.

7.3 Drivers of Forest Loss in Hamren, Rani-Garbhanga and Namdapha

In the Hamren sub-division of Karbi Anglong jhum, extraction of bamboo for sale at the Nagaon Paper Mill (NPM) and timber logging are proximate drivers of forest loss. Encroachment occurs only in selected areas such as the Amreng Reserved Forest (RF). Elsewhere in Hamren the land ownership rests with the community and the question of encroachment does not arise. Conversion of dense and open forests into grasslands has occurred due to the practice of jhum, while in the eastern part of the sub-division the practice of settled agriculture has affected the forest cover. In Hamren as in other hill areas the Sixth Schedule of the Indian Constitution entrusts ownership of land and forest resources to the community. Powers are vested in the District Councils and only limited forest areas are under state government (forest department) control. The Hamren example indicates that community ownership of forests as an institutional arrangement has not proved superior to that of state managed forests.

A slightly different set of drivers are at play in the Rani-Garbhanga RF. Being a RF the control is under the state's forest department; however, it is alleged that contractors and timber mafia operate with impunity (Assam Tribune 2012). On the southern fringe along the Assam-Meghalaya border illegal saw mills are in operation processing extracted timber. At a lower scale forest encroachment by small agriculturalists adds to the forest loss equation. Finally proximity to the million plus city of Guwahati on the northern limits of the RF has resulted in the spillover of urban population and an increase in the built-up area in the form of settlements (commercial as well as residential) and roads has occurred.

Like Rani-Garbhanga, the Namdapha National Park (NNP) is also under the control of the state forest department. Forest loss and degradation here is impacted upon by shifting cultivation and hunting traditionally practised by village tribesmen in the 25 odd villages within the park.

Thus a diverse sets of proximate and underlying drivers of forest loss are at play in the NER. The region's growing population has not helped as rising rural human population densities with low income levels have no option than to use timber resources as fuelwood. Steeped in poverty, marginal encroachers often take recourse to easy pickings in extracting small amounts of forest products for sale at roadside markets. With growing human population such individual effects add up to alarming proportions.

References

Angelsen A (1995) Shifting cultivation and "deforestation": a study from Indonesia. World Dev 23:1723–1729. doi:10.1016/0305-750X(95)00070-S

Anon (2002) Assam: protecting forest lands. Econ Pol Weekly 37:2516–2517

Assam Tribune (2012) Unabated timber felling on at Kamrup East. 30 November 2012, Guwahati. http://www.assamtribune.com/scripts/detailsnew.asp?id=nov3012/state05. Accessed 7 February 2013

Barthakur D (1977) Jhooming and its consequences. Yojana 21:122–123

Brown D, Schreckenberg K (1998) Shifting cultivators as agents of forest degradation: assessing the evidence. Natural resource perspectives 29. Overseas Development Institute, London

Das MM (2006) Illegal encroachment on 18,640 hectares. The Assam Tribune, Guwahati, 16 September 2006

Fox J (2000) How blaming 'slash and burn' farmers is deforesting mainland Southeast Asia. Asia Pacific Issues 47:1–8

Levy A, Scott-Clark C (2007) Poaching for Bin Laden, 5 May 2007. The Guardian, London. http://www.guardian.co.uk/world/2007/may/05/terrorism.animalwelfare. Accessed 25 May 2013

Myers N (1992) Tropical forests: the policy challenge. Environmentalist 12:15–27. doi:10.1007/BF01267592

Nongbri T (2001) Timber ban in North-East India: effects on livelihood and gender. Econ Pol Weekly 36:1893–1900

Rajimwale A (1991) Man verses ecology. Peoples Publishing House, New Delhi

Ramakrishnan PS (1985) Tribal man in the humid tropics of the north-east India. Man in India 65:1–32

Ranjan R, Upadhyay VP (1999) Ecological problems due to shifting cultivation. Current Science 77:1246–1250

Roy PS, Joshi PK (2002) Forest cover assessment in north-east India—the potential of temporal wide swath satellite sensor data (IRS-1C WiFS). Int J Remote Sens 23:4881–4896. doi:10.1080/01431160110114475

Saikia A (1998) Shifting cultivation, population and sustainability: the changing context of Northeast India. Development 41:97–100

Saikia A (2008) Forest fragmentation in North East India. In: Deka S (ed) North East India: geo-environmental issues. Eastern Book House, Guwahati

Saikia A, Hazarika R, Sahariah D (2013) Land use land cover change and fragmentation in the Nameri Tiger Reserve, India. Geografisk Tidsskrift (Danish J Geogr) 113:1–10. doi:10.1080/00167223.2013.782991

Sarma S (1987) Shifting cultivation in Meghalaya. North Eastern Geogr 19:61–69

SFR (1991) State of forest report 1989. Forest survey of India. Ministry of Environment and Forests, Dehra Dun

SFR (1997) State of forest report 1995. Forest survey of India. Ministry of Environment and Forests, Dehra Dun

Singh AP (1990) Labour management relations in the plywood industry in Assam. Ph.D. thesis (unpublished), Gauhati University, Guwahati

Chapter 8
Conclusions

Abstract Human populations have transformed ecosystems locally, regionally and globally. In the north east India biodiversity hotspot as well, human populations are slowly transforming the region's ecosystems. There can be no room for complacency taking the convenient but untenable explanation that other biodiversity hotspots are equally under threat. In many ways north east India represents a scenario wherein the progression and acceleration of the anthropocene has occurred. The region is emblematic of the challenges faced by societies that are barely industrialized and heavily forest sector dependent. Population growth has rendered age old practices like shifting cultivation (jhum) untenable and progressively less ecologically friendly. If the anthropocene is conceptualized as denoting the current interval of time, dominated by human activity then India's north eastern region fits in very precisely, given its systematic domination of the landscape by humans. High human population density in areas close to protected areas and the biomass requirements of these households is problematic in north east India. Fuelwood, issues of encroachment and illegal logging are all eating into the vitals of the region's forest resources. Indeed the time is ripe to act to save the rich forest resources of this hotspot from further decimation.

Keywords Anthropocene • Human dominated landscape • Forest • Hotspot • Shifting cultivation

8.1 Human Domination in the Anthropocene

North east India (NER) is a biodiversity hotspot with a high degree of endemic species. Unfortunately its forest wealth has been at the short end of the stick since the past half a century (Tucker 1988a, b; Saikia 1998). Are the forests of the NER uniquely imperiled? Probably not, since forest loss is occurring throughout the world (in various other hotspots as well) and we are currently in the anthropocene period of human existence (Tickell 2011). The term 'Anthropocene' was initially coined by Paul Crutzen (Slaughter Slaughter 2012); it is an informal term used widely and increasingly to signal the impact of collective human activity on bio-logical, physical and chemical processes on the Earth system (Zalasiewicz et al. 2011). The term anthropocene "was coined at a time of dawning realization that human activity was indeed changing the Earth on a scale comparable with some of the major events of the ancient past" and some of these changes are now seen as permanent, even on a geological time-scale (Zalasiewicz et al. 2010; Ellis 2011).

There is widespread realization among Earth and environmental scientists as evi-denced from scientific literature that some anthropogenic changes are comparable with those of the great forces of Nature (Zalasiewicz et al. 2011; Steffen et al. 2011a).

In many ways north east India represents a scenario wherein the progression and acceleration of the anthropocene has occurred. The region is emblematic of the challenges faced by societies that are barely industrialized and heavily forest sector dependent. Population growth has rendered age old practices like shifting cultiva-tion (jhum) untenable and progressively less ecologically friendly. If the anthropo-cene is conceptualized as denoting the current interval of time, dominated by human activity (Zalasiewicz et al. 2010), then India's north eastern region fits in very pre-cisely, given its systematic domination of the landscape by humans. All this with barely much industrial activity and barely any metropolitan city. While one of the most plainly visible physical effects of the anthropocene on the landscape has been the growth of the world's megacities (Zalasiewicz et al. 2010), the NER is marked by the complete absence of any megacity. In fact there is only a single million city, Guwahati in the area but the impress of forest extraction activities has been perva-sive. Human populations and their use of land have already transformed most of the terrestrial biosphere directly (Ellis 2011) and in the case of the NER only the north-ern parts of Arunachal Pradesh relatively free from such impact directly.

The human dominated landscape (Vitousek et al. 1997) overwhelmingly pre-vails in the NER. Forest resources are not always for local consumption and not insubstantial quantities exited the region, some to ports cities for shipment outside India as late as the early 1990s. An order by the Supreme Court, the apex court of India, banning the flow of timber resources from the region (Nongbri 2001) did not have the desired results. In less accessible forest areas, the Assam–Meghalaya border for instance, on the fringe of the Rani Reserved Forest, more than a dozen saw mills abound, and process timber with impunity. The quantum of such pro-duce is difficult to quantify, given that such operations are covert and illegal. However their cumulative effects coupled with other human stressors (jhum culti-vation and fuelwood extraction) leave visible results: a diminishing and degraded forest cover. The fact that government and community protected forests have been detrimentally affected means that governmentality levels have left much to be desired. In other countries as well weak governance and poor relations with nearby communities has undermined protected area effectiveness (Vuohelainen et al. 2012) and the NER fits into such a mould. Often the protectors of the forests (the respective state government's forest departments) themselves have been found to have been involved in clearing forests; although it needs to be pointed out that such instances reflect the activities of errant officials and hardly the entire departments per se. At the same time the fact remains that enforcement of forest protection levels are not at the desired level. Reports of encroachment in protected areas and illegal extraction of forest resources and well as of unfettered poaching of rare species continue to be frequently reported in the press and media. Unfortunately wildlife protection and protection of wildlife habitats and protected forest areas does not figure very high on the priority list of state agenda. All over the world ecosystems have been rapidly transformed in the post-2000 period by human populations through increasingly permanent uses of land (Ellis et al. 2010). Deforestation refers to the complete destruction of forest cover resulting in the forest biomass being so severely depleted that the remnant ecosystem is a travesty of natural forest as properly understood (Myers 1991). Various forms of agriculture and over-logging can result in the decline of biomass and depletion of ecosystem services in such a severe manner that the residual forest can no longer qualify as a forest (Myers 1991) and this is precisely what is happening in the tropical forests of north east India. Although efforts have been made to minimize shifting cultivation (Tisdell and Roy 1997) the results have not been successful enough; nor has been the spatial extent of areas weaned away from shifting cultivation been extensive. Most tropical forests are extraordinarily rich in species, remarkably complex ecologically and disappearing at truly alarming rates (Laurance 2007; Myers 1991). Generally extinction rates are likely to be higher in biodiversity hotspots which are geographically restricted areas with high species endemism, heavy habitat loss and rapidly increasing human populations (Laurance 2007). Although the NER is not a geographically restricted area comparable to Madagascar, the Brazilian Atlantic forests or the Philippines, it is an area with high species endemism that has suffered heavy habitat loss in the face of rapidly increasing human population. Its biodiversity is under increasing threat in many of its protected areas, reflective of poor enforcement and protection levels.

Starting with fairly transient practices like hunting and gathering and progressing to the increasingly permanent use of land for agriculture and settlements, human populations have transformed ecosystems locally, regionally and globally (Ellis et al. 2010). In the NER as well human populations are slowly transforming the region's ecosystems and there can be no room for complacency taking the excuse that other biodiversity hotspots are equally under threat.

8.2 Hotspot Under Duress

While it is true that "The world's most remarkable places are also the most threatened. These are the Hotspots: the richest and most threatened reservoirs of plant and animal life on Earth." (Conservational International 2013), at the same time much uncertainty exists about the nature and magnitude of anthropogenic impacts on tropical forest organisms (Laurance et al. 2011). The uncertainty arises from variability of the projected effects of climate change on tropical forests and tropical ecosystems and their vulnerability to changes in thermal and precipitation regimes (Laurance et al. 2011). As to what degree tropical forests such as those of north east India will be affected by the climatic element remains unclear; what is beyond dispute, however, is the anthropogenic dimension. Population growth, rising population densities and their living space are incrementally impinging on the tropical ecosystem and the region's biodiversity. Within the tropics, Southeast Asia has the highest rates of forest loss and degradation (Achard et al. 2002). As in the rest of India where high human population density exists in areas close to protected areas and the biomass requirements of these households is an agent of degradation of forests and loss of biodiversity (Davidar et al. 2010); the fuelwood issue for economically marginalized populations is problematic in the NER. Fuelwood consumption and biomass requirements are an important cause of forest decline in many developing countries (Bhatt and Sachan 2004; Davidar et al. 2007; Puyravaud et al. 2010; Rawat et al. 2009) and India and north east India is no different. What is pertinent, however, is that even in some protected areas in the NER, such as the Nameri National Park fuelwood extraction (along with other causes) effects forest degradation and forest loss (Saikia et al. 2013). Protection measures, with community participation, must be sought urgently and best practices from other tropical contexts must be adapted and adopted within a better late than never time frame. In fact the urgency of nature protection in the NER has long been stated (Tisdell and Roy 1997) and can only be reiterated here. Previous studies have pointed out that biota in tropical regions will be affected by far more than rising temperatures per se as habitat loss and fragmentation and rising hunting pressure exacerbate the effects of rising temperature (Laurance et al. 2011) and hitherto less experienced events such as tropical droughts may intensify and become more frequent (Phillips et al. 2009). Indeed committed and sustained efforts are expeditiously needed given that the synergistic set of issues (Laurance et al. 2011; Brodie et al. 2012) confronting the NER's tropical forest ecosystem. It is widely accepted that widespread biodiversity losses could affect the regulating services of the Earth System (Steffen et al. 2011a, b) and that even small changes in the extent of tropical forests can have impacts on climate, biodiversity and human wellbeing (Banfai and Bowman 2007). It is time that India looks more keenly to protect its biodiversity resources and hotspots and considers the issues beyond the confines of north east India.

References

Achard F, Eva HD, Stibig H, Mayaux P, Gallego J, Richards T, Malingreau J (2002) Determination of deforestation rates of the world's humid tropical forests. Science 297:999–1002. doi:10.1126/science.1070656

Banfai DS, Bowman DMJS (2007) Drivers of rain-forest boundary dynamics in Kakadu National Park, northern Australia, a field assessment. J Trop Ecol 23:73–86. doi:10.1017/S0266467406003701

Bhatt BP, Sachan MS (2004) Firewood consumption pattern of different tribal communities in northeast India. Energy Policy 32:1–6. doi:10.1016/S0301-4215(02)00237-9

Brodie J, Post E, Laurance WF (2012) Climate change and tropical biodiversity: a new focus. Trends Ecol Evol 27:145–150. doi:10.1016/j.tree.2011.09.008

Conservational International (2013) The biodiversity hotspots. http://www.conservation.org/where/priority_areas/hotspots/Pages/hotspots_main.aspx. Accessed 24 May 2013

Davidar P, Arjunan M, Mammen PC, Garrigues JP, Puyravaud JP, Roessingh K (2007) Forest degradation in the Western Ghats biodiversity hotspot: resource collection, livelihood concerns and sustainability. Curr Sci 93:1573–1578

Davidar P, Sahoo S, Mammen PC, Acharya P, Puyravaud JP, Arjunan M, Garrigues JP, Roessingh K (2010) Assessing the extent and causes of forest degradation in India: where do we stand. Biol Conserv 143:2937–2944. doi:10.1016/j.biocon.2010.04.032

Ellis EC, Goldewijk KK, Siebert S, Lightman D, Ramankutty N (2010) Anthropogenic transformation of the biomes 1700–2000. Glob Ecol Biogeogr 19:589–606. doi:10.1111/j.1466-8238.2010.00540.x

Ellis EC (2011) Anthropogenic transformation of the terrestrial biosphere. Phil Trans R Soc. Am 369:1010–1035. doi:10.1098/rsta.2010.0331

Laurance WF (2007) Have we overstated the tropical biodiversity crisis. Trends Ecol Evol 22:65–70. doi:10.1016/j.tree.2006.09.014

Laurance WF, Useche DC, Shoo LP, Herzog SK, Kessler M, Escobar F, Brehm G, Axmacher JC, Chen IC, Gámez LA, Hietz P, Fiedler K, Pyrcz T, Wolf J, Merkord CL, Cardelus C, Marshall AR, Ah-Peng C, Aplet GH, Arizmendi MD, Baker WJ, Barone J, Brühl CA, Bussmannx RW, Cicuzzae D, Eilu G, Favila MF, Hemp A, Hemp C, Homeier J, Hurtado J, Jankowski J, Kattána G, Kluge J, Krömer T, Lees DC, Marcus Lehnert M, Longino JT, Lovett J, Martin PH, Patterson BD, Pearson RG, Peh KSH, Richardson B, Richardson M, Samways MJ, Senbeta F, Smith TB, Utteridge TMA, Watkins JE, Wilson R, Williams SE, Thomas CD (2011) Global warming, elevational ranges and the vulnerability of tropical biota. Biol Conserv 144:548–557. doi:10.1016/j.biocon.2010.10.010

Myers N (1991) Tropical forests: present status and future outlook. Climatic Change 19:3–32. doi:10.1007/BF00142209

Nongbri T (2001) Timber ban in north-east india: effects on livelihood and gender. Econ and Polit Weekly 36:1893–1900

Puyravaud JP, Davidar P, Laurance WF (2010) Cryptic loss of India's native forests. Science 329:32. doi:10.1126/science.329.5987.32-b

Phillips OL, Aragao LEOC, Lewis SL, Fisher JB, Lloyd J, Lopez-Gonzalez G, Malhi Y, Monteagudo A, Peacock J, Quesada CA, van der Heijden G, Almeida S, Amaral I, Arroyo L, Aymard G, Baker TR, Banki O, Blanc L, Bonal D, Brando P, Chave J, de Oliveira ACA, Cardozo ND, Czimczik CI, Feldpausch TR, Freitas MA, Gloor E, Higuchi N, Jimenez E, Lloyd G, Meir P, Mendoza C, Morel A, Neill DA, Nepstad D, Patino S, Penuela MC, Prieto A, Ramirez F, Schwarz M, Silva J, Silveira M, Thomas AS, Steege H, Stropp J, Vasquez R, Zelazowski P, Davila EA, Andelman S, Andrade A, Chao KJ, Erwin T, Fiore AD, Honorio EC, Keeling H, Killeen TJ, Laurance WF, Cruz AP, Pitman NCA, Vargas PN, Ramirez-Angulo H, Rudas A, Salamao R, Silva N, Terborgh JW, Torres-Lezama A (2009) Drought sensitivity of the Amazon rainforest. Science 323:1344–1347. doi:10.1126/science.1164033

Rawat YS, Vishvakarma SCR, Todaria NP (2009) Fuel wood consumption pattern of tribal communities in cold desert of the Lahaul valley, north-western Himalaya, India. Biomass Bioenergy 33:1547–1557. doi:10.1016/j.biombioe.2009.07.019

Saikia A (1998) Shifting cultivation, population and sustainability: the changing context of north-east india. Development 41:97–100

Saikia A, Hazarika R, Sahariah D (2013) Land use land cover change and fragmentation in the Nameri Tiger Reserve, India. Geografisk Tidsskrift-Danish Journal of Geography 113:1–10. doi:10.1080/00167223.2013.782991

Slaughter RA (2012) Welcome to the anthropocene. Futures 44:119–126. doi:10.1016/j.futures.2011.09.004

Steffen W, Persson A, Deutsch A, Zalasiewicz J, Williams M, Richardson K, Crumley C, Crutzen P, Folke C, Gordon L, Molina M, Ramanathan V, Rockstrom J, Scheffer M, Schellnhuber HJ, Svedin U (2011a) The Anthropocene: from global change to planetary stewardship. Ambio 40:739–761. doi:10.1007/s13280-011-0185-x

Steffen W, Grinevald J, Crutzen P, McNeill J (2011b) The Anthropocene: conceptual and historical perspectives. Phil Trans R Soc Am 369:842–867. doi:10.1098/rsta.2010.0327

Tucker RP (1988a) The depletion of India's forests under British imperialism: planters, foresters, and peasants in Assam and Kerala. In: Worster D (ed) The ends of the earth: perspectives on modern environmental history. Cambridge University Press, Cambridge

Tucker RP (1988b) The British empire and India's forest resources: the timberlands of Assam and Kumaon 1914–1950. In: Richards JF, Tucker RP (eds) World deforestation in the twentieth century. Duke University Press, Durham

Tickell C (2011) Societal responses to the Anthropocene. Phil Trans R Soc Am 369:926–932. doi: 10.1098/rsta.2010.0302

Tisdell C, Roy K (1997) Sustainability of land use in north-east India: Issues involving economics, the environment and biodiversity. International Journal of Soc Econ 24:160–177. doi:10.1108/03068299710161188

Vitousek PM, Mooney HA, Lubchenko J, Melillo JM (1997) Human domination of earth's ecosystems. Science 227:494–499. doi:10.1126/science.277.5325.494

Vuohelainen AJ, Coad L, Marthews TR, Malhi Y, Killeen TJ (2012) The effectiveness of contrasting protected areas in preventing deforestation in Madre de Dios, Peru Environ Manage 50:645–663. doi:10.1007/s00267-012-9901-y

Zalasiewicz J, Williams M, Steffen W, Crutzen P (2010) The new world of the Anthropocene. Environ Sci Technol 44:2228–2231. doi:10.1021/es903118j

Zalasiewicz J, Williams M, Fortey R, Smith A, Barry TL, Coe AL, Bown PR, Rawson PF, Gale A, Gibbard P, Gregory FJ, Hounslow MW, Kerr AC, Pearson P, Knox R, Powell J, Waters C, Marshall J, Oates M, Stone P (2011) Stratigraphy of the Anthropocene. Phil Trans R Soc Am 369:1036–1055. doi:10.1098/rsta.2010.03151471-2962